Lab Manual to Accompany

Sixth Edition

# AGRISCIENCE

## FUNDAMENTALS AND APPLICATIONS

Lab Manual to Accompany

Sixth Edition

# AGRISCIENCE
## FUNDAMENTALS AND APPLICATIONS

Renée Peugh
Walter York

CENGAGE Learning

Australia • Brazil • Mexico • Singapore • United Kingdom • United States

CENGAGE Learning

**Lab Manual to Accompany *Agriscience: Fundamentals and Applications*, 6th Edition**
**Renée Peugh**
**Walter York**

General Manager: Dawn Gerrain
Associate Product Manager: Nicole Sgueglia
Senior Director, Development: Marah Bellegarde
Senior Product Development Manager: Larry Main
Senior Content Developer: Laura J. Stewart
Product Assistant: Scott Royael
Executive Director of Marketing: Carolyn Pestritto
Sr. Product Marketing Manager: Leah Klein
Senior Production Director: Wendy A. Troeger
Production Manager: Mark Bernard
Senior Content Project Manager: Elizabeth C. Hough
Senior Art Director: David Arsenault
Cover image(s): Field: © tlorna/Shutterstock.com; Cover and Section Opener image credits: (top image) Windmill: © Michal Bednarek/Shutterstock.com; (second image) Sheep: © iStockphoto/Joan Wynn; (third image) Water with Boat: © Vladislav Gajic/Shutterstock.com; (bottom image) Plants: © David Kay/Shutterstock.com



Library of Congress Control Number: 2013943109

ISBN-13: 978-1-133-68689-7

**Cengage Learning**
200 First Stamford Place, 4th Floor
Stamford, CT 06902
USA

Cengage Learning is a leading provider of customized learning solutions with office locations around the globe, including Singapore, the United Kingdom, Australia, Mexico, Brazil, and Japan. Locate your local office at: **www.cengage.com/global**

Cengage Learning products are represented in Canada by Nelson Education, Ltd.

To learn more about Cengage Learning, visit **www.cengage.com**

Purchase any of our products at your local college store or at our preferred online store **www.cengagebrain.com**



Printed in the United States of America
1 2 3 4 5 6 7 18 17 16 15 14

# CONTENTS


Preface ........ vii

## SECTION 1 Agriscience in the Information Age / 1

Exercise 1 There's a Method to the Madness—The Scientific Method ........ 2
Exercise 2 Getting Up Close and Personal—The Microscope ........ 5
Exercise 3 It's Alive!—The Cell ........ 10
Exercise 4 Don't Drink the Water: An Insight into Water Safety ........ 15
Exercise 5 How Degrading—Maintaining a Healthy Environment ........ 18

## SECTION 2 You and the New Millennium / 23

Exercise 6 Get a Job—Your Career in Agriscience ........ 24
Exercise 7 Experience Required—Your SAE (Supervised Agricultural Experience) ........ 36
Exercise 8 Show Me the Way—Leadership in Agriscience ........ 41

## SECTION 3 Natural Resources Management / 47

Exercise 9 Watt's the Big Deal—Creating Electricity Out of Thin Air ........ 48
Exercise 10 Getting Down and Dirty—Preventing Soil Erosion ........ 51
Exercise 11 Talking Dirty—Soil Fertility ........ 55
Exercise 12 Hold the Soil Please—Growing Plants without Soil ........ 59
Exercise 13 There's Something Fishy Here—Fish Anatomy ........ 62

## SECTION 4 Integrated Pest Management / 71

Exercise 14 Are You a Competitor? Weed Identification Lab ........ 72
Exercise 15 Don't Bug Me—Insect Anatomy ........ 74
Exercise 16 Stuck on You—External Parasites ........ 78

## SECTION 5 Plant Sciences / 81

Exercise 17 Putting Down Some Roots—Root Anatomy ........ 82
Exercise 18 It Stems from—Stem Anatomy ........ 86
Exercise 19 Turn Over a New Leaf—Leaf Anatomy ........ 91
Exercise 20 He Loves Me, He Loves Me Not—Flower Anatomy ........ 95
Exercise 21 Under the Knife—Flower Dissection ........ 99
Exercise 22 Free Gas Made Here—Photosynthesis ........ 102
Exercise 23 Transpiration—Nature's Big Gulp ........ 105
Exercise 24 Reap What You Sow—Seed Anatomy ........ 107

Exercise 25 Come Out; Come Out Wherever You Are—Germination.............. 110
Exercise 26 Let's See What You're Made of—DNA Extraction..................... 115

SECTION 6 Crop Science / 119

Exercise 27 We Are Here, We Are Here, We Are Here—Soil Organisms............ 120
Exercise 28 A Good Start—Growing and Transplanting Vegetable Seedlings............................................. 124
Exercise 29 Try That On for Size—Grafting Fruit Trees........................... 128
Exercise 30 From Fries to Fuel—Making Biodiesel................................ 131
Exercise 31 Deep Thoughts—Exploring Seed Planting Depths .................. 134
Exercise 32 Does This Smell Good?—Storing Silage ............................ 137
Exercise 33 Stop Buggin' Me—Insect Collection ................................ 140

SECTION 7 Ornamental Use of Plants / 143

Exercise 34 Keep 'em in the Dark—Plants and Light ............................ 144
Exercise 35 Chip Off the Old Block—Plant Cloning ............................. 146

SECTION 8 Animal Sciences / 149

Exercise 36 You Eat Like a Pig—Simple Animal Digestion........................ 150
Exercise 37 It Takes Four—Ruminant Digestion ................................. 154
Exercise 38 Must Have Been Something I Ate—Internal Parasites................ 159
Exercise 39 We Are Doing What?—Fecal Examination for Parasites ............. 164
Exercise 40 It's All in the Genes—Understanding Genetics...................... 167
Exercise 41 This Will Hurt Only for a Minute—Injection Procedures ............. 171
Exercise 42 Hot Wings Anyone?—Animal Anatomy and Physiology ............. 174

SECTION 9 Food Science and Technology / 179

Exercise 43 Break It on Down?—Food Nutrients................................. 180
Exercise 44 What's for Dinner?—Exploring Choose My Plate..................... 184
Exercise 45 Is This Still Good?—Food Preservation.............................. 186
Exercise 46 How Cheesy—Dairy Product Production ............................ 188
Exercise 47 Does This Milk Make Me Look Fat?—Homogenization .............. 190
Exercise 48 Do You Have a Sweet Tooth?—Fruit Sugars ......................... 192

SECTION 10 Communications and Management in Agriscience / 195

Exercise 49 Where Did the Money Go?—Understanding Diminishing Returns.............................................. 196
Exercise 50 I Owe You—Interest and Credit ..................................... 201
Exercise 51 In the Money—Writing a Business Plan.............................. 205
Exercise 52 It's a Balancing Act—Balance Sheets ................................ 207
Exercise 53 Rich Man, Poor Man—Profit-Loss Statements ...................... 211
Exercise 54 Money, Money, Money—Interest ..................................... 215
Exercise 55 Can't We All Just Get Along—Cooperative Math Simulations.............................................. 218

# PREFACE

This laboratory manual was developed to accompany *Agriscience: Fundamentals & Applications*, Sixth edition. The topics and exercises throughout were designed to meet the science-based curriculum needs of the Agriscience 1 course in secondary schools. Although written with the organization of *Agriscience: Fundamentals & Applications*, Sixth edition, in mind, this can be used as a stand-alone laboratory manual.

New to this edition of the lab manual, optional Internet supplements offer additional research opportunities and educational resources to learn more about topics covered in the lab exercises. Each lab exercise has been enhanced with new photos and illustrations to help stimulate visual learning. For example, Lab Exercise #22 on photosynthesis has been updated to provide students with a better method of data collection and analysis. Exercise #44 has been updated to reflect the new USDA's Choose My Plate initiative.

## ABOUT THE AUTHOR

Reneé Peugh is a graduate of Boise State University. While there, she studied Biology and Secondary Education. She is a certified teacher with endorsements in biology, the earth sciences, chemistry, and health science. She has served as the biological consultant and contributing author for the text that accompanies this manual: *Agriscience Fundamentals and Applications, Introduction to Forestry Science* and other commercial projects. Renee understands the value of agricultural education and the FFA program. She served as a chapter and Idaho State FFA officer and is currently a 4-H leader.

## EQUIPMENT AND SUPPLIES LIST

The equipment and supplies listed here were carefully selected to be used with the exercises in the *Lab Manual to Accompany Agriscience: Fundamentals & Applications,* Fifth edition. All items listed here are available from Carolina Biological Supply Company, 2700 York Road, Burlington, NC 27215. You may also place your order or request a catalog by calling toll free 1-800-334-5551 or by logging onto www.carolina.com.

| | Carolina Biological Supply Catalog Number |
|---|---|
| **Exercise 2** | |
| Compound microscope | 590950 |
| Beginner's slide set, which includes a letter "e," newsprint, colored threads, cork, cornstarch, salt crystals, dust, cotton fibers, nylon fibers, silk fibers, wool fibers, and volcanic ash slides | 291030 |
| **Exercise 3** | |
| Compound microscope | 590950 |
| Animal cell slide | 312250 |
| Plant cell slide | 301700 |
| Chromosomes—Mitosis slide | 308946 |

| | Carolina Biological Supply Catalog Number |
|---|---|
| **Exercise 3 (Continued)** | |
| Onion epidermis slide | 303662 |
| Glass microscope slides | 632950 |
| Coverslips | 972729 |
| Medicine droppers | |
| **Exercise 4** | |
| Pond water | |
| Medicine droppers | 736903 |
| Glass microscope slides and coverslips | 632950 |
| Microscopes with 30X and 100X power | 632950 & 972729<br>590950 |
| **Exercise 5** | |
| Paper strips | |
| Dried plant materials such as leaves | |
| Scissors | |
| Spray bottle | |
| Water | |
| Three plastic cups | |
| Sand | |
| Topsoil | |
| Plastic | |
| Biodegradable plastic | |
| Cotton fibers | |
| Synthetic fibers | |
| Magnifying glass | 953801 |
| Gloves | 706501 |
| **Exercise 6** | |
| Thumbtack | |
| Sheet of posterboard—18 inches × 24 inches | |
| Scissors | |
| Glue | |
| *Occupational Outlook Handbook* | |
| *Dictionary of Occupational Titles* | |
| **Exercise 7** | |
| *Official FFA Manual* | |
| SAEP Record Book | |

| | Carolina Biological Supply Catalog Number |
|---|---|
| **Exercise 8** | |
| *Official FFA Manual* | |
| *Parliamentary Guide for FFA* or | |
| *Robert's Rules of Order* | |
| **Exercise 9** | |
| 100 cm of copper wire | |
| 1 bar magnet | |
| 1 electrical meter | |
| 1 cardboard tube | |
| Duct tape | |
| Fan or turn table | |
| **Exercise 10** | |
| Six plant trays (or shoe boxes cut to 5 inches deep) | |
| Plastic sheeting to line trays | |
| Six sprinkling cans | |
| Six beakers | 721210A |
| Trowel | |
| Scissors | |
| Stopwatches | |
| Soil | |
| Sod (grass-covered soil from a lawn) | |
| Water | |
| Supports to form inclines for trays | |
| **Exercise 11** | |
| Seven identical 6-inch flower | |
| Pots with drainage outlets at the bottom | |
| Pot chip or screen | |
| Vermiculite or perlite | |
| Mixing bin or box | |
| Pea and tomato seeds | |
| Plant Macronutrient | |
| Kit that includes the following; | 653201 |
| Nitrogen source—$NH_4NO_3$ | |
| Phosphorus source—superphosphate | |

| | Carolina Biological Supply Catalog Number |
|---|---|
| **Exercise 11(Continued)** | |
| Potassium source—KCI | |
| Calcium source—$CaCO_3$ | |
| Materials for labeling pots | |
| Laboratory balance or scale | |
| Ruler or yardstick | |
| **Exercise 12** | |
| Personal safety set | |
| Goggles | 972160 |
| Apron | 706240 |
| Gloves | 706336 |
| Textbook | |
| Dissection | |
| Lab Perch Kit | 221580 |
| One pair of scissors | |
| One scalpel | 626520 |
| One sharp probe | 627370 |
| Magnifying glass | 953801 |
| Personal safety set | |
| Goggles | 972160 |
| Apron | 706240 |
| Gloves | 706336 |
| **Exercise 14** | |
| A large amount of newspaper | |
| Two pieces of tempered hard board, peg board or plywood (~12" ×18" rectangles) | |
| Cardboard into ~12" ×18" rectangles | |
| Tie downs or bungee cords for strapping the weed portfolio tightly | |
| Large pieces of construction paper for mounting the specimens | |
| Plastic bags for collecting (old grocery bags work well and keep the specimens moist) | |
| Shovel | |
| Adhesive for mounting the specimens | |
| Blank adhesive labels | |
| 20 weeds collected near your home | |

| | Carolina Biological Supply Catalog Number |
|---|---|
| Weed identification manuals | |
| Small zip style bags | |
| **Exercise 15** | |
| Dissection manual | 521004 |
| Grasshopper kit | 221430 |
| One pair of fine dissection scissors | |
| One sharp probe | 627370 |
| Dissecting microscope | 593006 |
| Personal safety set | |
| Goggles | 972160 |
| Apron | 706240 |
| Gloves | 706336 |
| **Exercise 16** | |
| Compound microscope (used for most lice and fleas) | 590950 |
| Dissecting scope (sufficient for larger lice and engorged ticks) | 593006 |
| Insecticide powder | |
| Animal comb or any inexpensive plastic hair comb | |
| Smooth dark material (paper or plastic) | |
| 70 percent ethyl alcohol or isopropyl (rubbing) alcohol | |
| Glass jar | |
| Water or oil (baby oil) | |
| Microscope slides and cover slips | |
| Dissecting needle | 627220 |
| Personal safety set | |
| Goggles | 972160 |
| Apron | 706240 |
| Gloves | 706336 |
| **Exercise 17** | |
| Monocot and dicot root slide set | 301898 |
| Compound microscope | 590950 |
| Set of colored pencils | |

| | Carolina Biological Supply Catalog Number |
|---|---|
| **Exercise 18** | |
| Monocot and dicot stem slide set | 302642 |
| Compound microscope | 590950 |
| Set of colored pencils | |
| **Exercise 19** | |
| Leaf section slide | 304066 |
| Compound microscope | 590950 |
| Set of colored pencils | |
| **Exercise 20** | |
| *Taraxacum* young bud slide | 304382 |
| *Taraxacum* older bud slide | |
| Mature anther slide | 304586 |
| Pollen germination slide | 304270 |
| Compound microscope | 590950 |
| Set of colored pencils | |
| **Exercise 21** | |
| A complete flower | |
| Dissecting probe (teasing needles) | 627370 |
| Scissors | |
| Scalpel (optional) | 625924 |
| Tweezers/forceps (optional) | |
| Dissecting scope or zoom lens | 593006 |
| 18 to 108 mm | |
| **Exercise 22** | |
| Calomba plant or other water plant | |
| Test tube | 731048 |
| Test tube clamp | 702902 |
| Water | |
| Rectangular beaker | |
| High-intensity grow light | |
| **Exercise 23** | |
| 2 geranium leaves | |
| 2 small jars (baby food or | |
| Mason jars, or 2 beakers) | |
| Cardboard (10 cm square with a hole about the size of a hole-punch in center) | |
| Clay or petroleum jelly | |

| | Carolina Biological Supply Catalog Number |
|---|---|
| **Exercise 24** | |
| Seed of a plant slide | |
| Compound microscope | 590950 |
| Several fresh bean seeds | |
| Magnifying lens | 953801 |
| Scalpel | 625924 |
| **Exercise 25** | |
| Water | |
| Aluminum foil | |
| Beaker | 721210A |
| Hot plate | 701002 |
| Freezer | |
| Seed-Gro Kit | |
| **Exercise 26** | |
| 1 plastic quart-size bag | |
| 1 strawberry 10 mL DNA extraction buffer | |
| Filter paper | 712800 |
| Ice cold 90 percent ethanol or isopropyl alcohol | |
| 1 Test tube or champagne flute | |
| Funnel (optional) | 653590 |
| Inoculation loop or coffee stir straw | 703039 |
| Glass microscope slides | |
| Coverslips | |
| What's in the Soil II— | |
| Microorganism kit | |
| Compound microscope—300X or better | 590950 |
| Four 400-mL beakers | 721210A |
| Test tube rack | 731880 |
| Two test tubes | 653590 |
| Bunsen burner | 706709 |
| Handful of green leaves | |
| Soil samples | |
| Water | |
| Personal safety set | |
| Goggles | 972160 |
| Apron | 706240 |
| Gloves | 706336 |

| | Carolina Biological Supply Catalog Number |
|---|---|
| **Exercise 28** | |
| Tomato seeds | |
| Seed-starting media | |
| Flats | |
| Spray bottle with water | |
| Trowel or spoon | |
| Plastic wrap | |
| Large containers (10 oz. or larger) | |
| Plant-growing media | |
| Liquid fertilizer | |
| Ruler | |
| Toothpicks | |
| **Exercise 29** | |
| Knife | |
| Grafting wax | |
| Grafting tape | |
| Apple or other fruit stock—¼- to ½-inch diameter | |
| Scions—¼- to ½-inch diameter | |
| **Exercise 30** | |
| 1 liter *new* vegetable oil | |
| 4 grams substance A | |
| 250 mL substance B | |
| 2 glass containers with lids (glass juice bottle) | |
| 1 funnel | |
| 4 qt pot | |
| Warming plate | 701002 |
| Thermometer | 745513 |
| Blender (Do not use with food again) | |
| Turkey baster (Do not use with food again) | |
| **Exercise 31** | |
| Six glass jars—pint size | |
| Soil | |
| Water | |
| Corn seeds | |

| | Carolina Biological Supply Catalog Number |
|---|---|
| **Exercise 32** | |
| Freshly cut chopped grass | |
| Six 1-quart jars with tight-fitting lids | |
| Labels for the jars | |
| Pen | |
| Nails | |
| Hammer | |
| **Exercise 33** | |
| Hand lens 10–30X power, or a dissecting scope | 593006 |
| Sweep net (can be purchased from any biological educational catalog) | |
| Stapler | |
| Paper bags (large grocery store size) | |
| Glass jars with screw-top lids | |
| Ethanol or isopropyl alcohol (rubbing alcohol) | |
| Use of a freezer (optional) | |
| Glass dish or Petri dish | |
| **Exercise 34** | |
| Six pots of pea seedlings | |
| Four light-tight boxes, each with a hole in one side | |
| One completely light-tight box Light source | |
| **Exercise 35** | |
| Kit | |
| African violet plant | |
| Soap | |
| Bunsen burner | |
| Scalpel | 625924 |
| Water—Sterile | |
| Four to five large beakers | 721210A |
| African violet potting soil | |
| Ethyl alcohol | |
| Hot plate | |
| Laundry bleach | |
| Ten to twelve small plastic pots | |

| | Carolina Biological Supply Catalog Number |
|---|---|
| **Exercise 35 (Continued)** | |
| Personal safety set | |
| Goggles | 972160 |
| Apron | 706240 |
| Gloves | 706336 |
| **Exercise 36** | |
| Textbook | |
| Scalpel | 625924 |
| Dissecting microscope | 593006 |
| Personal safety set | |
| Goggles | 972160 |
| Apron | 706240 |
| Gloves | 706336 |
| Preserved digestive tract of a pig or a fetal pig | |
| **Exercise 37** | |
| Textbook | |
| Digestive tract of a pig or fetal pig | 228391 |
| Rubber gloves | 706336 |
| Scalpel | 625924 |
| Four jars | |
| Dissecting microscope | 593006 |
| Compound microscope | 590950 |
| Rumen bacteria slide | |
| Personal safety set | |
| Preserved digestive tract of a cow or sheep | |
| **Exercise 38** | |
| Sheep liver | |
| Fluke microscope | |
| Slide set | 306406 |
| Sheep tapeworm | |
| Microscope | |
| Slide set | 306766 |
| Pig roundworm | |
| Microscope slide | 307020 |
| Compound microscope | 590950 |

| | Carolina Biological Supply Catalog Number |
|---|---|
| **Exercise 39** | |
| Compound microscope | 590950 |
| Plastic bag (Ziploc® type) | |
| Jelly bean size of fresh feces of a cat, dog, sheep, cattle, hog, or horse | |
| Microscope slide | 632950 |
| Coverslip | 972729 |
| Small piece of newspaper | |
| Toothpick | |
| Tap water | |
| Eyedropper | 736913 |
| Personal safety set | |
| Goggles | 972160 |
| Apron | 706240 |
| Gloves | 706336 |
| **Exercise 40** | |
| Pencil or pen | |
| Calculator | |
| Text | |
| Internet access (optional) | |
| **Exercise 41** | |
| Two or three 10-mL syringes | |
| ½-inch needles | |
| 1-inch needles | |
| Three containers of water | |
| Food dye - Red, green, blue | |
| Three or four grapefruit | |
| Ruler | |
| **Exercise 42** | |
| One raw chicken wing for each student. (Chicken wings can be purchased at the local supermarket in the inexpensive family pack.) | |
| Scissors | |
| Paper plate | |
| Tweezers or forceps | |
| Protective gloves | 706336 |

| | Carolina Biological Supply Catalog Number |
|---|---|
| **Exercise 43** | |
| Twenty test tubes | |
| Test tube rack | |
| Five eyedroppers | 736913 |
| Hot plate | 701002 |
| Sudan IV | 892993 |
| Biuret reagent | 848209 |
| Potassium iodide | 872793 |
| Benedict's solution | 847121 |
| Milk | |
| Cream or butter | |
| Syrup | |
| Canned whole corn | |
| Distilled water | |
| **Exercise 44** | |
| Variety of foods from each food group | |
| Kitchen measuring tools | |
| Scale in ounces | |
| **Exercise 45** | |
| Bread | |
| Swiss cheese | |
| Hot plate | 701002 |
| 20-mL test tubes | 731009 |
| 250-mL beaker | 721175 |
| Distilled water | |
| Test tube holder | 732105 |
| Sulfuric acid solution | 893431 |
| Funnel | |
| Filter paper | |
| Spatula | |
| Ethyl alcohol | |
| **Exercise 46** | |
| 12-inch piece of cheesecloth | |
| Hot plate | 701002 |

| | Carolina Biological Supply Catalog Number |
|---|---|
| Plastic cups—16 ounces each | |
| Cooking thermometer | |
| Milk | |
| Salt | |
| Rennin or rennilase | |
| **Exercise 47** | |
| Compound microscope with an oil immersion lens | 590950 |
| Glass microscope slides | |
| Coverslips | |
| Health food store that carries raw milk | |
| Homogenized milk | |
| **Exercise 48** | |
| Refractometer | |
| A variety of fruits | |
| Kitchen knife | |
| Blender (optional) | |
| **Exercise 49** | |
| Calculator | |
| **Exercise 50** | |
| Calculator | |
| **Exercise 51** | |
| Computer | |
| Calculator | |
| **Exercise 52** | |
| Calculator | |
| **Exercise 53** | |
| Calculator | |
| **Exercise 54** | |
| Calculator | |

# SECTION 1

# Agriscience in the Information Age

Courtesy of DeVere Burton

# EXERCISE 1

## There's a Method to the Madness—The Scientific Method

### PURPOSE

"The scientific method" is best described as the steps scientists and students follow when trying to explain the world around them. These procedures were used by the scientists who discovered how to add beta-carotene to rice. Golden rice has blessed many people in underdeveloped countries where a lack of beta-carotene was causing blindness and poor immunity. In this exercise, you will learn and use the scientific method.

### MATERIALS NEEDED

- Textbook
- Other materials deemed appropriate by students

### PROCEDURE

Study the steps of the scientific method below. These steps will help you throughout this laboratory manual.

**Step 1.** Through observation, identify the problem. Ask the question: "What, exactly, do I want to find out?" Limit your topic to a single research objective.

**Step 2.** Review the literature. Read up on and become familiar with the topic.

**Step 3.** Form a hypothesis. Develop a hypothesis or statement to be proven or disproven.

**Step 4.** Prepare a project proposal. Outline how the project should be done. Include timelines, facilities, equipment required, costs, and a description of how you will do the project.

**Step 5.** Design the experiment. Develop a plan for carrying out the project.

**Step 6.** Collect the data. Conduct the experiment. Record what you measure or observe.

**Step 7.** Draw conclusions. Summarize the results. Make calculations. Determine if the information collected allows you to accept or reject the hypothesis.

**Step 8.** Prepare a written report. This report should include your hypothesis, a brief project description, results, and the conclusions. This report will provide you and others with a permanent record of your research.

## OBSERVATIONS

Using the scientific method in agriscience

**Step 1.** Identify the problem. The following problem has been identified for you. Complete the remaining steps of the scientific method.

How does the color of a material affect the heat and light absorption or reflection?

______________________________

______________________________

______________________________

**Step 2.** Review the literature. In the space provided, list any references you used for review. Include the name of the references and other pertinent information that will help you find the resources again (example, Web address).

______________________________

______________________________

______________________________

**Step 3.** Form a hypothesis. Write your hypothesis in the space provided.

______________________________

______________________________

______________________________

**Step 4.** Prepare a project proposal. Outline your project in the space provided. Include timelines, facilities, equipment required, costs, and a description of how you will do the project.

______________________________

______________________________

______________________________

**Step 5.** Design the experiment. Develop your plan to test your hypothesis in the space provided.

______________________________

______________________________

______________________________

**Step 6.** Collect the data. Record any data in the space provided. Design your own data format.

______________________________

______________________________

______________________________

**Step 7.** Draw conclusions. Summarize your results in the space provided. Accept or reject your hypothesis.

**Step 8.** Prepare a written report. Using a computer and word processing program, prepare a written report not to exceed two pages explaining your project and the results.

## CONCLUSIONS

1. How did the scientific method help you to prove or disprove your hypothesis?

2. Which step in the scientific method do you consider to be the most important and why?

3. Which step in the scientific method do you consider to be the least important and why?

## OPTIONAL INTERNET SUPPLEMENT

Rice University, Center for Technology in Teaching and Learning
http://webadventures.rice.edu/. Search- "virtual clinical trials."
This Web page allows students to apply the scientific method.

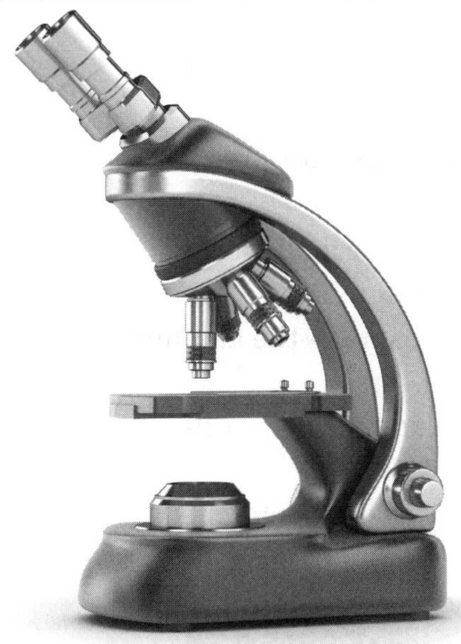

© Alex Mit/Shutterstock.com

# EXERCISE 2

## Getting Up Close and Personal—The Microscope

### PURPOSE

The microscope is an amazing tool that allows you to see objects too small to be seen with the naked eye. In this exercise, you will be introduced to the compound microscope. You will learn microscope anatomy, how to use one, and finally practice using it.

### MATERIALS NEEDED

- Compound Microscope
- Beginner's Slide Set, that includes a letter "e," newsprint, colored threads, cork, cornstarch, salt crystals, dust, cotton fibers, nylon fibers, silk fibers, wool fibers, and volcanic ash slides

## PROCEDURE

**1.** Study the diagram of a microscope. Learn the names and functions of the parts.

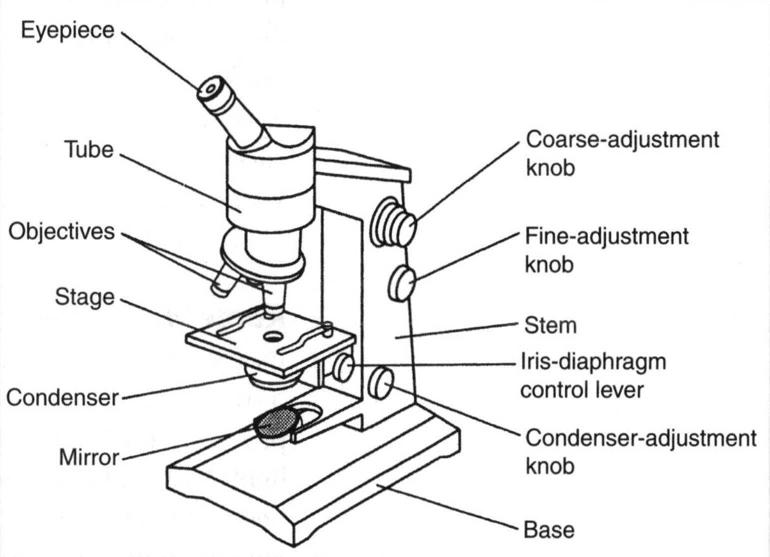




a. **eyepiece**—magnifying lens at the top of the microscope
b. **tube**—connects eyepiece to the objectives
c. **objectives**—lens that magnifies the object to be viewed
d. **stage**—platform that holds the slide for viewing
e. **condenser**—concentrates the light source on the object being observed
f. **light source**—provides light to view the slide on the stage; it may be a mirror or a lamp. This is commonly called the illuminator.

g. **coarse-adjustment knob**—knob used for initial focusing
h. **fine-adjustment knob**—knob used for final focusing
i. **stem**—connects the base to the lens assembly
j. **iris-diaphragm control lever**—controls the amount of light to the slide
k. **condenser-adjustment knob**—adjusts light brilliance through the condenser
l. **base**—platform to which entire microscope is attached; sits on table or desk

**2.** Prepare the microscope and view the slide.

a. If your microscope has a mirror, position the mirror so light is reflected through the hole on the stage. If your microscope has a light, turn it on.
b. Place the letter "e" microscope slide on the stage and secure it with the stage clips. Make sure the coverslip faces up. Center the slide over the stage hole.
c. Rotate the low-power objective so it locks into position directly over the specimen and over the stage hole.
d. Look at the stage and use the coarse-adjustment knob to bring the objective down toward the specimen, being careful not to touch the lens with the coverslip.
e. While observing through the eyepiece, rotate the coarse-adjustment knob counterclockwise to bring the image into view. Use the fine-adjustment knob for sharp focusing. ***Remember:*** *Always focus upward.*
f. Practice opening and closing the iris diaphragm by moving the lever on the side of the condenser clockwise to provide more light and counterclockwise to provide less light. More light is needed for higher-power objectives. Less light is required for transparent objects. Record your observations in Observation 2.
g. While looking through the eyepiece, observe the letter "e" slide while moving the slide forward, backward, left, and right. Describe what happens to the "e" in Observation 3.
h. Carefully turn the objective to the high-power lens. Look through the eyepiece at the letter "e" slide. If it is not in focus, adjust the fine-adjustment knob only. (Never use the coarse-adjustment knob to adjust the high-power objective lens because this lens is very close to the slide and could break it.)
i. Observe each of the other slides found in the beginner's slide set. Draw a picture of or describe what you see for each slide in Observation 4.

## OBSERVATIONS

**1.** Label the parts of the microscope in the following diagram.

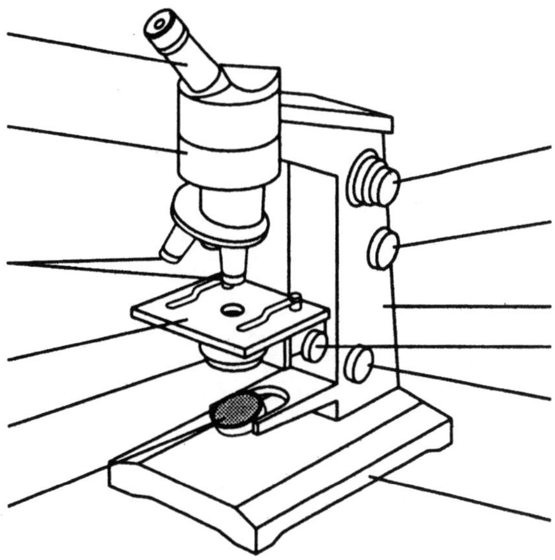



**2.** Describe the effects of opening and closing the diaphragm.

______________________________________________

______________________________________________

______________________________________________

**3.** Describe the appearance of the letter "e" under the microscope and how it moves as the slide moves.

______________________________________________

______________________________________________

______________________________________________

**4.** Draw a picture of or describe in detail each of the following as it appears under the microscope.

a. newspaper print

b. colored threads

c. cork

d. cornstarch

e. salt crystals

f. dust

g. cotton fibers

h. nylon fibers

i. silk fibers

j. wool fibers

k. volcanic ash

## CONCLUSIONS

1. How does the image you see through the microscope appear when compared with the non-magnified object seen directly?

2. Explain how the movement of the slide on the stage and the movement as viewed through the microscope differ.

3. In what areas of agricultural research would the microscope be useful? List at least five.

## OPTIONAL INTERNET SUPPLEMENT

University of Delaware, VD Virtual Compound Microscope simulation
www.udel.edu/biology/ketcham/microscope/scope.html
This Web page offers a simulation of using a microscope.

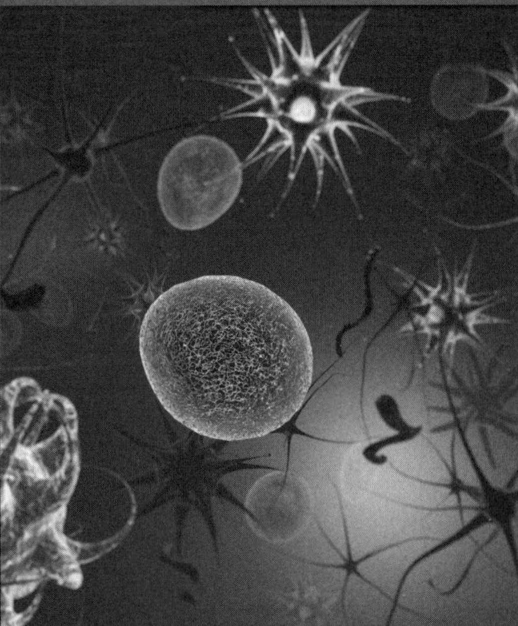



# EXERCISE 3

## It's Alive!—The Cell

### PURPOSE

This exercise allows you to study and compare the structures of plant and animal cells and identify the common parts of cells.

### MATERIALS NEEDED

- Compound microscope
- Animal cell slide
- Plant cell slide
- Chromosomes—Mitosis slide
- Mitochondria slide
- Golgi apparatus slide
- Onion epidermis slide
- Glass microscope slides
- Coverslips
- Medicine droppers
- Cell reference materials

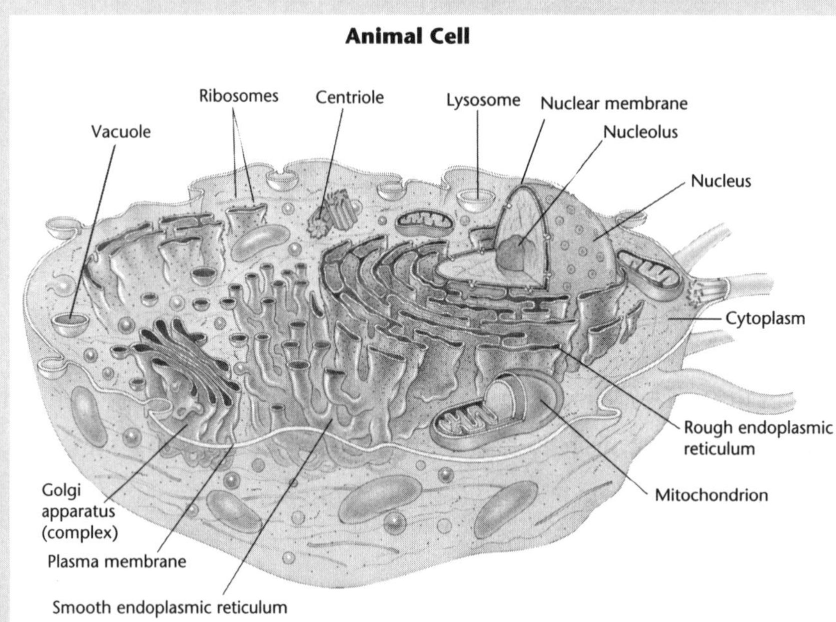




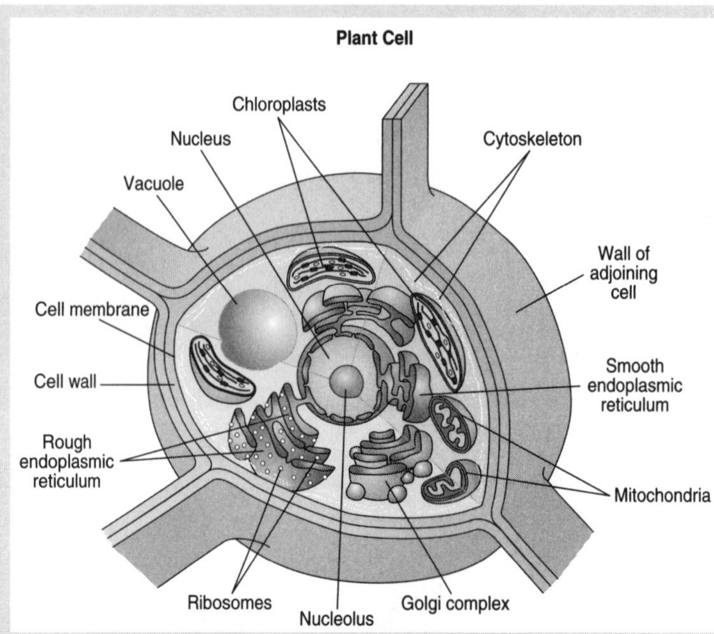

## PROCEDURE

1. Viewing animal cells
   a. Place the general animal cell slide on the stage of the microscope and view it under low power.
   b. Switch to high power. In Observation 1 that follows, identify these parts: cell membrane, cytoplasm, nucleus, nucleolus, and nuclear envelope.
   c. Place the chromosome slide on the stage of the microscope and view it under low power.
   d. Switch to high power. In Observation 2, identify the chromosomes in the drawing.
   e. Place the mitochondria slide on the stage of the microscope and view it under low power.
   f. Switch to high power. In Observation 3, identify the mitochondria in the drawing.
   g. Place the Golgi apparatus slide on the stage of the microscope and view it under low power.
   h. Switch to high power. In Observation 4, identify the Golgi apparatus in the drawing.
   i. Use the general animal cell picture to help you complete the statements in Observation 5.
2. Viewing plant cells
   a. Place the general plant cell slide on the stage of the microscope and view it under low power.
   b. Switch to high power. In Observation 6, identify the following parts: cell wall, plasma membrane, cytoplasm, nucleus, central vacuole, chloroplasts, and plasmodesmata.
   c. Use the general plant cell picture to help you complete the statements in Observation 7.
3. Viewing onion cells
   a. Place the onion epidermis slide on the stage of the microscope and view it under low power.
   b. Switch to high power. In Observation 8, draw an onion cell and identify the following parts: cell wall, plasma membrane, cytoplasm, nucleus, and vacuoles.

## OBSERVATIONS

**1.** Identify the parts of the general animal cell on the following diagram.

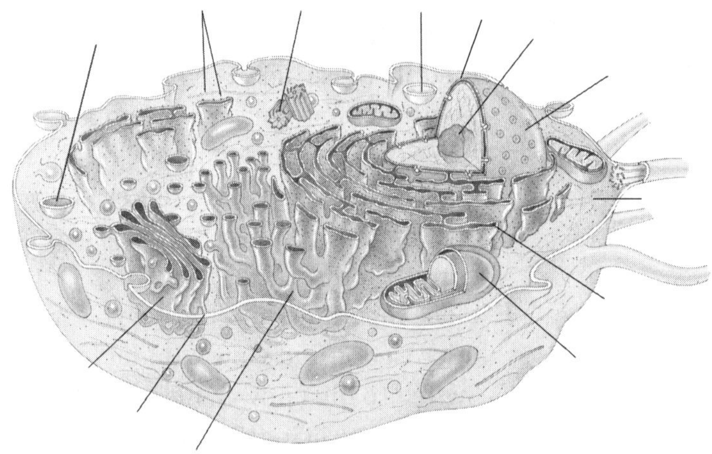



**2.** Using the Internet and any other cell reference materials provided by your instructor, fill in the blanks to make the following statements true:

a. The _______________ _______________ is the outermost structure of an animal cell.

b. The fluid medium in which much of the intermediary metabolism of the cell takes place is the _______________.

c. The large organelle that contains the cell's inherited information and serves to control and coordinate all metabolism is the _______________.

d. The _______________ separates the nuclear contents from the rest of the cell.

e. The _______________ are tangled strands that extend throughout the nucleus and contain genes.

f. Cellular organelles in which the chemical energy in food molecules is converted to a form useful to cellular metabolism are called _______________ .

g. The _______________ plays a part in the storage or secretion of metabolic products.

h. The site of ribosomal RNA synthesis and ribosome assembly is the _______________.

**3.** Identify the parts of the general plant cell on the following diagram.

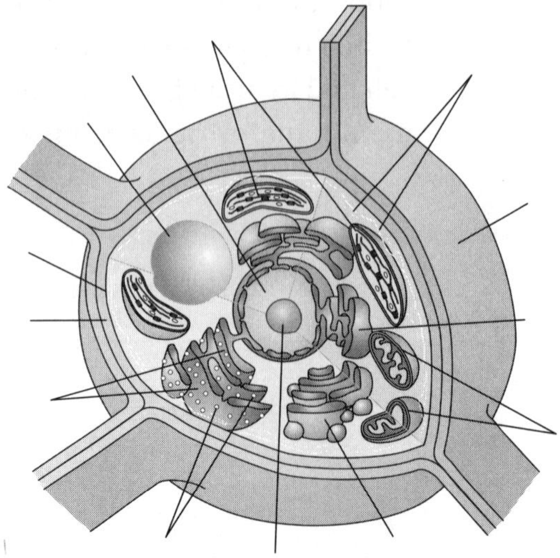

**4.** Fill in the blanks to make the following statements true.

a. The ______________ ______________ is the outermost structure of a plant cell.

b. The membrane-enclosed sac that occupies up to 90 percent of the cell volume is the ______________.

c. The large organelle that contains the cell's inherited information and serves to control and coordinate all metabolism is the ______________.

d. The ______________ consists of a viscous fluid matrix that contains water, many organic and inorganic substances, and a variety of tiny, specialized, functional structures.

e. The ______________ forms the outer boundary of a cell and maintains the chemical composition of the cell by regulating the passage of substances in and out of the cell.

f. ______________ are organelles that often contain lamella; chloroplasts are an example.

**5.** Draw an onion cell and label its parts.

## CONCLUSIONS

**1.** What parts are different between animal and plant cells?

______________________________________________

______________________________________________

______________________________________________

**2.** What are the differences between the general plant cell and the onion cell?

______________________________________________

______________________________________________

______________________________________________

**3.** What are some other plant cells that would be different from the general plant cell?

______________________________________________

______________________________________________

______________________________________________

4. What parts of the general animal and plant cells are similar?

5. What differences can be observed in the general shapes of the animal and plant cells?

## OPTIONAL INTERNET SUPPLEMENT

The Biology Corner (www.biologycorner.com) offers additional activities and resources related to plant and animal cells, among other topics. Search "plant cells" or "animal cells."

# EXERCISE 4

## Don't Drink the Water: An Insight into Water Safety

© Don Farrall/Photodisc/Getty Images

### PURPOSE

In many parts of the world people do not have access to clean potable (drinkable) water. In more developed countries, water purification systems are a vital part of daily living. After completing this exercise, you will have a greater appreciation for these systems. You will also be able to observe and identify some small organisms that live in still water.

### MATERIALS NEEDED

- Pond water
- Medicine droppers
- Glass microscope slides and coverslips
- Microscopes with 30× and 100× power
- Internet access or a pond water organism field reference book

### PROCEDURE

1. Each student will collect a small jar of water from a local pond. Allow for the water to sit in the class room near a window for a few days prior to the lab for incubation. Note: river and stream water should not be used; they will not produce favorable results.
2. Place one drop of water onto the slide. Place your coverslip onto the slide by holding the coverslip at a 45° angle to the water then gently drop it onto the slide.
3. Examine the slide under the microscope at 30×. When you have found an organism, adjust the microscope to 100×.

Using the Internet or reference books, try to identify your organism. To find a good identification key on the Internet, use keywords "pond water key." Guides may also be found under the "Images" browsers using key words "pond water organisms."

## OBSERVATIONS

1. Draw a picture of three organisms you found.

2. How does each organism move?

3. Identify any organelles you may be able to see, and label them on your original drawing.

## CONCLUSIONS

1. What are the names of the organisms you found?

2. Do you think these organisms are single celled or multicellular? Why?

3. Compare and contrast the organisms you found.

4. Do you think the organisms you found may be harmful to humans?

5. On a scale from 1 to 10, how important are water purification systems? Why?

## OPTIONAL INTERNET SUPPLEMENT

Microbus is a Web site that shows common microscopic organisms found in unclean water (www.microscope-microscope.org). Navigate to Applications, Pond Water Critters: A guide to protozoans and microscopic animals.

USDA/ARS # K4175-13

# EXERCISE 5

## How Degrading—Maintaining a Healthy Environment

### PURPOSE

In this exercise you will determine the solid-waste materials that are biodegradable and explore some of the factors that influence the rate of degradation.

### MATERIALS NEEDED

- Land Fills and Environmental Kit
- Paper strips
- Dried plant materials such as leaves
- Scissors
- Spray bottle
- Water
- Three plastic cups
- Sand
- Topsoil
- Plastic
- Biodegradable plastic
- Cotton fibers
- Synthetic fibers
- Magnifying glass
- Gloves

### PROCEDURE

***Note:*** Students may work individually or in small groups.

**1.** Preparing samples
   a. Cut the paper into ½-by-2-inch strips. Cut the two kinds of plastic into ½-by-2-inch strips.
   b. Cut the synthetic and natural fibers into 2-inch lengths.
   c. Cut the dried plant materials into pieces about 2 inches long and no wider than ½ of an inch. This can be done in a food processer.
   d. Separate the materials into three piles. Each pile should have equal parts of all of the materials being tested.

**2.** Preparing the landfill
   a. Label three plastic cups as follows:
      Cup #1—Sand
      Cup #2—Dry topsoil
      Cup #3—Moist topsoil
   b. Fill cup #1 with sand, up to 1 inch from the top, packing the sand tightly.
   c. Fill cups #2 and #3 with topsoil, up to 1 inch from the top, packing the soil lightly.
   d. Water the topsoil in cup #3 using a spray bottle. The soil should be uniformly moist and not have any residual water pooled at the bottom of the cup.

**3.** Adding materials to the landfill
   a. Make initial observations of the original condition of the following materials to be tested; record your observations in Table 5-1.

**Name** ______________________ **Date** ______________

**Table 5-1.** **Cup #** ______________ **Soil Type** ______________

| Material | General Appearance and Condition | Color | Texture |
|---|---|---|---|
| Paper | | | |
| | | | |
| | | | |
| | | | |
| Plastic | | | |
| | | | |
| | | | |
| | | | |
| Biodegradable plastic | | | |
| | | | |
| | | | |
| | | | |
| Natural fiber | | | |
| | | | |
| | | | |
| | | | |
| Synthetic fiber | | | |
| | | | |
| | | | |
| | | | |
| Dried plant material | | | |
| | | | |
| | | | |
| | | | |

**Name** ______________________________ **Date** ______________

**Table 5-2.** **Cup #** ______________ **Soil Type** ______________

| Fiber | At Start | First Week | Second Week | Third Week | Fourth Week | Fifth Week | Sixth Week | Seventh Week |
|---|---|---|---|---|---|---|---|---|
| Natural (Cotton) | +++ | | | | | | | |
| Synthetic (Polyester) | +++ | | | | | | | |

Rate fiber strength as follows:

+++ Original strength at start

++ Starting to show signs of disintegration, such as breaks or cuts of the thinner fibers that make up the sample

+ Very weak, but still held together

– Disintegrated or decomposed



1. paper strips
2. plastic strips
3. biodegradable plastic strips
4. natural fibers
5. synthetic fibers
6. dried plant material

*Note:* Other materials can also be tested for biodegradation.

b. Test the original strength of the fiber (synthetic and natural) by stretching the fiber; record fiber strength using the key provided in Table 5-2.

c. Insert a piece of each of the materials to be tested into the soil of each cup, covering about half the surface area of each material. A portion of each material tested should be exposed above the soil surface. The exposed portions can be used to locate the buried materials and as reference guides to the original conditions of the materials.
d. Press the materials lightly into the soil and cover each cup with a clear plastic top.
e. Place your cups on a tray and store them at room temperature.
f. Every few days, check material for moisture. The material should be kept slightly moist. If water is needed, lightly spray the composting material with a spray bottle. Too much water will result in failure.
g. The mixture needs to be stirred every few days to keep the process going.

## OBSERVATIONS

1. Record your observations of the original condition of the materials in Table 5-1. ***Note:*** Ask your instructor for multiple copies of Table 5-1 for use throughout the experiment.
2. Record your observations of the original strength of the fibers in Table 5-2.
3. Observe your setup once a week and record your results in Table 5-1. Remove the materials one by one, gently clean off the soil material, and examine each carefully with a 10X magnifying glass. Look for differences in general characteristics, such as size, shape, texture, breaks in the fibers, and fading of colors.
4. Test the strength of the fibers (synthetic and natural) once a week and compare these results to those of the original fibers. Record your observations in Table 5-2.

## CONCLUSIONS

1. Which of the soil materials provided for the best biodegradation? Explain your answer.

______________________________________________

______________________________________________

______________________________________________

2. When did you first notice signs of degradation?

______________________________________________

______________________________________________

______________________________________________

**3.** What difference was there in the degradation between the natural fiber and the synthetic fiber?

______________________________

______________________________

______________________________

**4.** Which of the cups had the most degrading activity? Explain why you think it did.

______________________________

______________________________

______________________________

**5.** Which material decomposed the fastest? Why do you think it had the highest rate of degradation?

______________________________

______________________________

______________________________

## OPTIONAL INTERNET SUPPLEMENT

University of Illinois Extension (http://web.extension.illinois.edu):
Search-Composting for the Homeowner.
This Web site provides a tutorial for small scale composting.

# SECTION 2

# You and the New Millennium

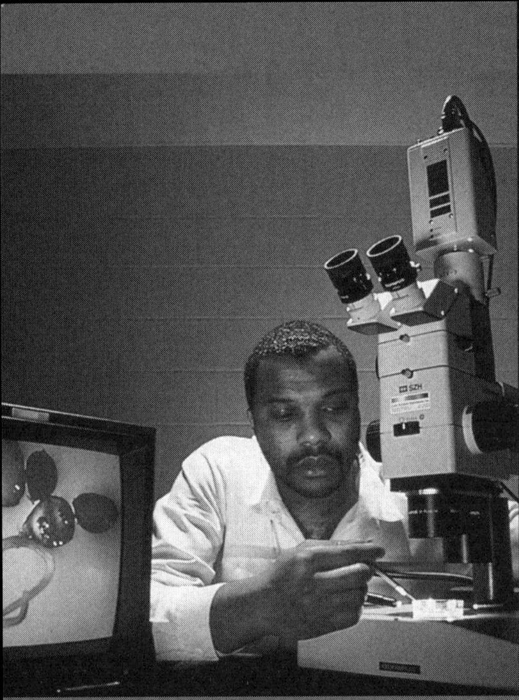

USDA/ARS #K4024-1

# EXERCISE 6

## Get a Job—Your Career in Agriscience

### PURPOSE

The one thing high school students have in common with each other is in a few short years every one of you should graduate and begin a career and/or seek advanced training to further your success. As adults, you will have financial obligations that require you to "get a job." This exercise will introduce you to a number of agriscience careers.

### MATERIALS NEEDED

- Pencil
- Pen or marker
- Thumb tack or paper fastener
- Sheet of poster board—18 inches × 24 inches
- Scissors
- Glue
- *Occupational Outlook Handbook*
- *Dictionary of Occupational Titles*
- Internet access

### PROCEDURE

**1.** Preparing the career wheel

a. Using a pair of scissors, cut out the career wheel.

b. Cut out a piece of poster board the same size as the career wheel.

c. Glue the career wheel to the cut out piece of poster board.

d. Cut a rectangular piece of poster board to measure 8½ inches by 11 inches.

e. Place the thumbtack in the middle of the career wheel and fasten it to the 8½-by-11-inch piece of poster board.

f. Cut out an arrow from the poster board. Place it on the 8½-by-11-inch piece of poster board pointing to the wheel.

g. When you finish, your career wheel should look like the diagram below.

Career Wheel
Agriscience Processing Products and Distribution
Horticulture
Agricultural Mechanics
Agriscience Professions
Agriscience Supplies and Services
Forestry
Renewable Natures Resources
Production Agriculture

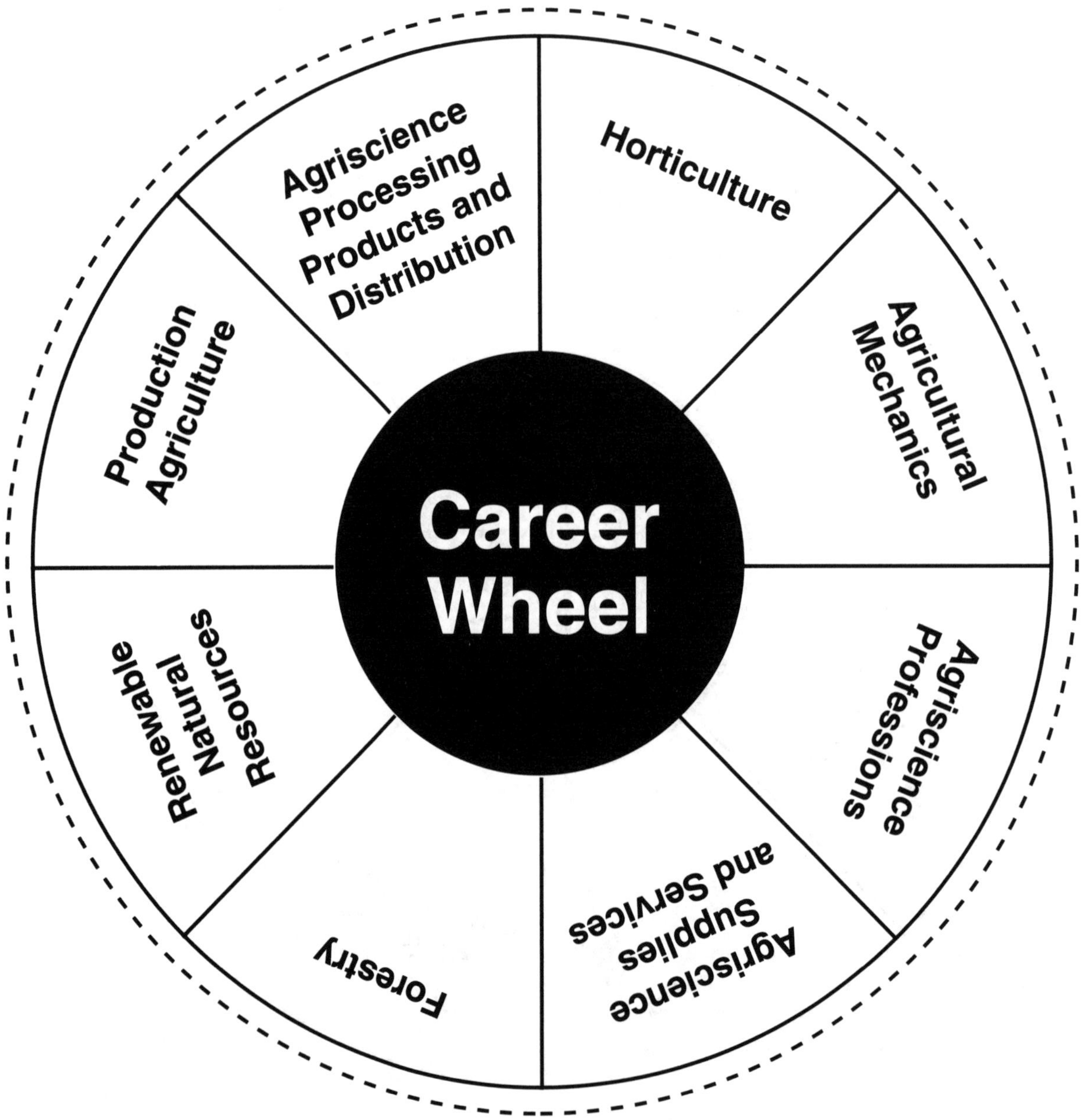
Career Wheel
Horticulture
Agricultural Mechanics
Agriscience Professions
Agriscience Supplies and Services
Forestry
Renewable Natural Resources
Production Agriculture
Agriscience Processing Products and Distribution

**2.** Choosing a career.
   a. Spin the career wheel. The arrow will point to one of the career areas.
   b. Select a job for the career area from the list found under Procedure 3.
   c. Using the *Occupational Outlook Handbook, Dictionary of Occupational Titles*, or the Internet, complete Observation 1 for the job you selected.
   d. Spin the wheel again and repeat procedures 2b and 2c to complete Observation 2.
   e. Without using the wheel, select a job of your choice from the lists under Procedure 3. Complete Observation 3 for the job you selected.

**3.** Agricultural career areas and jobs listed for each:
   a. Production Agriculture

| | |
|---|---|
| Crop Farmer | Dairy Farmer |
| Fruit and Nut Farmer | Livestock Producer |
| Horticultural Grower | Poultry Producer |
| Vegetable Farmer | Other List ____________ |

   b. Agriscience Processing, Products, and Distribution

| | |
|---|---|
| Ag Establishment Inspector | Butcher |
| Cattle Buyer | Christmas Tree Grader |
| Cotton Grader | Farm Stand Operator |
| Federal Grain Inspector | Food and Drug Inspector |
| Food Processing Supervisor | Fruit Distributor |
| Fruit and Vegetable Grader | Flower Grader |
| Fruit Press Operator | Grain Buyer |
| Grain Broker | Hog Buyer |
| Grain Elevator Operator | Livestock Yard Supervisor |
| Livestock Commission Agent | Meat Cutter |
| Meat Inspector | Produce Buyer |
| Milk Plant Supervisor | Quality Control Supervisor |
| Produce Commission Agent | Weights and Measures Official |
| Tobacco Buyer | Winery Supervisor |
| Other List ____________ | Wool Buyer |

   c. Horticulture

| | |
|---|---|
| Floral Designer | Floral Shop Operator |
| Florist | Golf Course Superintendent |
| Greenhouse Manager | Greens keeper |
| Horticulturist | Hydroponics Grower |
| Landscape Architect | Landscaper |
| Nursery Operator | Plant Breeder |
| Turf Farmer | Turf Manager |
| Other List ____________ | |

d. Forestry

| | |
|---|---|
| Forester | Forest Ranger |
| Heavy Equipment Operator | Log Grader |
| Logging Operations Inspector | Lumber Mill Operator |
| Nursery Operator | Park Ranger |
| Plant Breeder | Timber Manager |
| Tree Surgeon | Other List ____________ |

e. Renewable Natural Resources

| | |
|---|---|
| Animal Behaviorist | Animal Ecologist |
| Animal Taxonomist | Environmental Conservation |
| Environmentalist | Fire Warden |
| Forest Firefighter/Warden | Forest Ranger |
| Game Farm Supervisor | Game Warden Officer |
| Ground Water Geologist | Park Ranger |
| Range Conservationist | Resource Manager |
| Soil Conservationist | Trapper |
| Water Resources Manager | Wildlife Manager |
| Other List ____________ | |

f. Agriscience Supplies and Service

| | |
|---|---|
| Aerial Crop Duster | Ag Aviator |
| Ag Chemical Dealer | Agricultural Equipment |
| Ag Equipment Dealer | Animal Groomer |
| Animal Health Products | Animal Inspector |
| Animal Keeper | Animal Trainer |
| Artificial Breeding | Artificial Inseminator |
| Biostatistician | Chemical Applicator |
| Chemical Distributor | Computer Analyst |
| Computer Operator | Computer Programmer |
| Computer Salesperson | Crop Chemicals, Machinery |
| Custom Operator | Dairy Management Specialist |
| Dog Groomer | Farm Appraiser |
| Farm Auctioneer | Farrier |
| Feed Mill Operator | Feed Ration Developer |
| Fertilizer Plant Supervisor | Field Inspector and Analyst |
| Field Sales Representative | Fiber Technologist |
| Harness Maker | Harvest Contractor |
| Horse Trainer | Insect and Disease Inspector |
| Kennel Operator | Lab Technician |
| Meteorological Analyst | Pest Control Technician |
| Pet Shop Operator | Poultry Field Service |
| Poultry Hatchery Manager | Poultry Inseminator |
| Sales Manager | Salesperson |
| Service Technician | Sheep Shearer |
| Other List ____________ | |

g. Agricultural Mechanics
   - Ag Construction Engineer
   - Ag Electrician
   - Ag Equipment Designer
   - Ag Plumber
   - Ag Safety Engineer
   - Diesel Mechanic
   - Equipment Operator
   - Hydraulic Engineer
   - Irrigation Engineer
   - Land Surveyor
   - Machinist
   - Parts Manager
   - Research Engineer
   - Safety Inspector
   - Soil Engineer
   - Welder
   - Other List ______________

h. Agriscience Professions
   - Ag Accountant
   - Ag Advertising Executive
   - Ag Association Executive
   - Ag Consultant
   - Ag Corporation Executive
   - Ag Economist
   - Ag Educator
   - Ag Extension Educator
   - Ag Extension Specialist
   - Ag Journalist
   - Ag Lawyer
   - Ag Loan Officer
   - Ag Market Analyst
   - Ag Mechanics Teacher
   - Ag News Director
   - Agriculture Attaché
   - Agronomist
   - Animal Cytologist
   - Animal Geneticist
   - Animal Nutritionist
   - Animal Physiologist
   - Animal Scientist
   - Agriculturist
   - Avian Veterinarian
   - Bacteriologist
   - Biochemist
   - Bioengineer
   - Biophysicist
   - Botanist
   - Computer Specialist
   - Credit Analyst
   - Daily Nutrition Specialist
   - Dendrologist
   - Electronic Editor
   - Embryologist
   - Environmental Educator
   - Entomologist
   - Equine Dentist
   - 4-H Youth Assistant
   - Farm Appraiser
   - Farm Broadcaster
   - Farm Investment Manager
   - Food Chemist
   - Foreign Affairs Official
   - Graphic Designer
   - Herpetologist
   - Horticulture Instructor
   - Hydrologist
   - Ichthyologist
   - Information Director
   - International Specialist
   - Invertebrate Zoologist
   - Land Bank Branch Manager
   - Liminologist
   - Magazine Writer
   - Mammalogist
   - Marine Biologist
   - Marketing Analyst
   - Media Buyer
   - Microbiologist
   - Mycobiologist
   - Nematologist
   - Organic Chemist
   - Ornithologist
   - Ova Transplant Specialist
   - Paleobiologist
   - Parasitologist
   - Pharmaceutical Chemist
   - Photographer
   - Plant Cytologist
   - Plant Ecologist
   - Plant Geneticist
   - Plant Nutritionist
   - Plant Pathologist
   - Plant Taxonomist
   - Pomologist
   - Poultry Scientist
   - Public Relations Manager

| | |
|---|---|
| Publicist | Publisher |
| Reproductive Physiologist | Rural Sociologist |
| Satellite Technician | Scientific Artist |
| Scientific Writer | Silviculturist |
| Software Reviewer | Soil Scientist |
| Vertebrate Zoologist | Veterinarian |
| Veterinary Assistant | Veterinarian Pathologist |
| Veterinarian Technician | Virologist |
| Vidiculturist | Vocational Agriculture |
| Other List ____________ | Instructor/FFA Advisor |

## OBSERVATIONS

Find the information requested in Observations one through three by using the *Dictionary of Occupational Titles* and *Occupational Outlook Handbook.*

## CONCLUSIONS

1. How does the nature of work differ for each of the three jobs? How is it similar?

   ______________________________________________

   ______________________________________________

2. How do the working conditions differ for each of the three jobs? How are the conditions the same?

   ______________________________________________

   ______________________________________________

3. How do the qualifications for each of the three jobs differ? How are the qualifications the same?

   ______________________________________________

   ______________________________________________

4. How does the job outlook compare for each of the three jobs?

   ______________________________________________

   ______________________________________________

5. How do the earnings compare for each of the three jobs?

   ______________________________________________

   ______________________________________________

## OPTIONAL INTERNET SUPPLEMENT

The handbooks listed in the **Materials Needed** can be found at www.bls.gov/ooh/ (Bureau of Labor Statistics), and www.occupationalinfo.org.

## OBSERVATION 1

### Career Wheel Challenge

**Career Area** [ ] **Job Selected** [ ]

Nature of the work:

Working conditions:

Training and other qualifications:

Job outlook:

Earnings:

## OBSERVATION 2

### Career Wheel Challenge

**Career Area** ______________ **Job Selected** ______________

Nature of the work:

Working conditions:

Training and other qualifications:

Job outlook:

Earnings:

## OBSERVATION 3

### Career Wheel Challenge

**Career Area** ______________ **Job Selected** ______________

Nature of the work:

Working conditions:

Training and other qualifications:

Job outlook:

Earnings:

# Which Career Will Be a Winner for You!?

Name ______________________ Date ______________

Using your *Agriscience* textbook, turn to Figures 4-6, 4-8, 4-11, 4-13, 4-15, 4-17, 4-19, and 4-20 select one career that interests you and write that career on the bracket located on the left side of this activity sheet. (Choose two careers from Figure 4-20).

Make sure each career is different. Next, select one career from the top two left brackets and write the winner on the line connected to the top two brackets. Continue the process for the middle and bottom sets of brackets. Continue the same process until you have one career written on the single right bracket. That's YOUR winning career!!

Copyright © 2015 Cengage Learning®.

# EXERCISE 7

## Experience Required—Your SAE (Supervised Agricultural Experience)

Courtesy of The National FFA Organization

### PURPOSE

The purpose of this activity is to help plan a Supervised Agricultural Experience (SAE).

### MATERIALS NEEDED

- *Official FFA Manual*
- *SAE Record Book*
- Pen or pencil

### PROCEDURE

1. Planning your SAE
   a. For Observation 1, list the various agricultural crop or livestock enterprises produced successfully in your community. Circle those that interest you.
   b. For Observation 2, list the various agribusinesses found in your community that support production agriculture. Circle those that interest you.
   c. For Observation 3, list improvement activities you could carry out at your home, farm or in your community.
   d. List agriscience skills you have and could perform in Observation 4.
   e. Look through your *SAE Record Book*. What type of information would you record in it? Write your answer in Observation 5.
   f. Complete the Resources Inventory found in this exercise (Observation 6).
   g. Complete the "Selecting a Supervised Agricultural Experience Program" form found in this exercise (Observation 7).
   h. Optional: Visit the SAE of an outstanding agriscience junior or senior or an American FFA Degree recipient (if one is nearby). Ask how he or she got started.
2. SAE awards
   a. Use your *Official FFA Manual* to identify individual awards that you could apply for based on your proposed SAE. List these in the Conclusions section, question #1.
   b. Ask your agriscience instructor to help you identify other state and local awards for which you qualify. List those awards under Conclusions, question #2.

## OBSERVATIONS

*Refer to Procedure 1 in this exercise.*

1. Crop and livestock enterprises produced in my community:

2. Agribusinesses found in my community:

3. Improvement projects (applied activities) I can do:

4. Supplementary skills (applied activities) I can perform:

5. Information I would record in my SAE record book:

6. Complete the "Resources Inventory" form provided in this exercise.
7. Complete the "Selecting a Supervised Agricultural Experience Form" provided in this exercise.

## Resources Inventory

1. Name ______________________ Age ____________ Class ______________
2. Address ______________________ Phone ______________________
3. Parents' names ______________________________________
4. Occupations ______________________________________
5. Number of brothers and sisters ______________________
6. I live on a farm ____________ in town ____________ on an acreage ____________
7. Is land available for you to rent to produce crops? Yes ____________ No ____________
   a. If yes, how many acres? ______________ b. Which crops? ______________
   c. Location of the land ______________________________________
8. Are facilities available for you to rent to produce livestock or livestock products?

   Yes ______________ No ______________

   a. If yes, what type? ______________________________________
   b. Number ______________ c. Location of facilities ______________
9. Do you have space available for a garden? Yes __________ No __________
10. Do you have a greenhouse available for your use? Yes __________ No __________
11. Do you have facilities for doing mechanical work? Yes __________ No __________
12. Would you be interested in producing livestock on the school farm?

    Yes __________ No __________ If yes, what type of livestock? ______________
13. Would you be interested in producing crops on the school farm?

    Yes __________ No __________ If yes, what crop? ______________________
14. Do you have an agricultural job available to you?

    Yes __________ No __________ Describe ______________________

    ______________________________________________________

# SELECTING A SUPERVISED AGRICULTURAL EXPERIENCE

**for**

______________________________________

Name of Student

**Instructions:** Use this form to tentatively decide on a beginning agriscience SAE. This information will be used in agriscience classes to develop detailed plans for obtaining experiences.

My interest in agriscience is______________________________________

______________________________________

______________________________________

Based on my interest and the opportunities available to me to get practical experience in agriscience, I plan to include the following in my agriscience SAE:

1. Production or productive enterprises (examples: beef, dairy, nursery production, Christmas trees)

   ______________________________________

   ______________________________________

2. Placement in an agribusiness (examples: supply store, florist shop, nursery, golf course, landscape contracting)

   ______________________________________

   ______________________________________

3. Improvement (examples: landscape your home, fertilize your lawn, plant trees)

   ______________________________________

   ______________________________________

4. Agriscience skills (examples: change spark plugs, weld, change the oil in small engines)

   ______________________________________

   ______________________________________

5. Other activities (examples: projects on school facility)

   ______________________________________

   ______________________________________

## CONCLUSIONS

*Refer to Procedure 2 in this exercise.*

**1.** National individual awards for which I can apply. ______________________________

______________________________

______________________________

______________________________

Obtain the paperwork from your instructor, and apply for the awards you qualify for.

**2.** State or local awards for which I can apply. ______________________________

______________________________

______________________________

______________________________

Obtain the paperwork from your instructor, and apply for the awards you qualify for.

## OPTIONAL INTERNET SUPPLEMENT

The *Official FFA Manual* and *SAE* information can be found at the National FFA Organization's Web site, www.ffa.org. Search Resources.

Photo courtesy of The National FFA Organization

# EXERCISE 8

## Show Me the Way—Leadership in Agriscience

### MATERIALS NEEDED

- *Official FFA Manual*
- *Parliamentary Guide for FFA* or *Robert's Rules of Order*
- Pen or pencil

### PURPOSE

Exercise 8 will introduce you to the official procedures for conducting FFA meetings and the rules of parliamentary law.

### PROCEDURE

**1.** Conducting FFA meetings

a. Study the following diagram of the FFA meeting room. Learn where each of the officers is stationed and the symbol of each office.

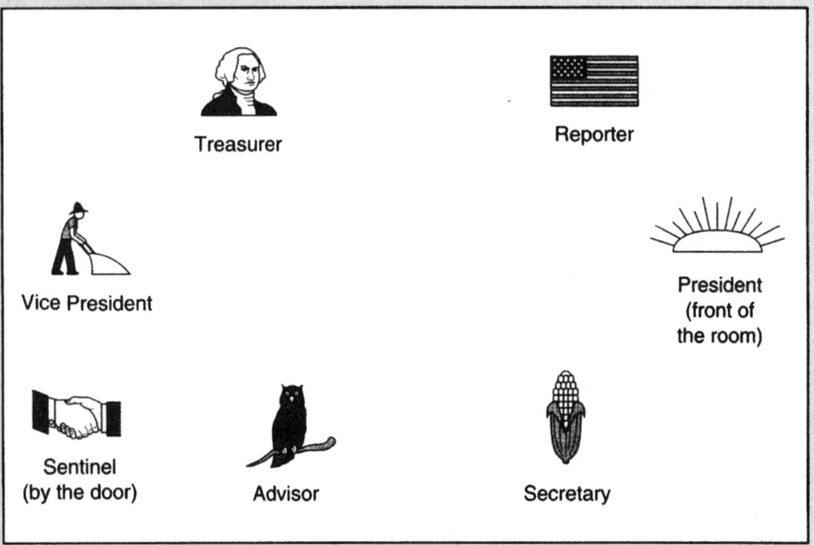




b. In the *Official FFA Manual*, read over the opening and closing ceremonies for all FFA meetings.

c. Let each class member read a part in the ceremonies. Take turns. Go over the ceremonies several times.

d. Select one officer's position, and write the office's speaking part in the ceremonies in Observation 2.

**2.** Learning parliamentary law

a. In the *Official FFA Manual*, study the order of business for a meeting. Complete Observation 3.

b. Study the classifications, names, purposes, and pertinent information for the primary motions used in conducting FFA business. Use the *Parliamentary Guide for FFA* or *Robert's Rules of Order*. Complete Observations 4 and 5.

c. Practice the rules of parliamentary law using the "Practicing Parliamentary Law" activity section found in this exercise. Fill in the blanks appropriately.

## OBSERVATIONS

*Use the FFA Manual to complete the following questions.*

**1.** Complete the following meeting room diagram by listing the officers and the symbols where they should be stationed.

Front of the room

Door

**2.** After reviewing the chapter FFA offices, consider what office you would like to run for and write that office's speaking part for FFA's opening ceremonies.

______________________________________________

______________________________________________

**3.** List the order of business for an FFA meeting.

______________________________________________

______________________________________________

**4.** Fill in the blanks for the chart below.

| Name of Motion | Kind of Motion | Debatable Yes or No | Amendable Yes or No | Vote Required Majority Two-thirds or None | Second Required Yes or No |
|---|---|---|---|---|---|
| Main Motion | | | | | |
| Lay on the Table | | | | | |
| Division of the House | | | | | |
| Amend | | | | | |
| Take from the Table | | | | | |
| Refer to a Committee | | | | | |
| Previous Question | | | | | |
| Reconsider | | | | | |
| Parliamentary Inquiry | | | | | |
| Rescind | | | | | |
| Suspend the Rules | | | | | |
| Adjourn | | | | | |

## PRACTICING PARLIAMENTARY LAW

1. A member moves to paint the interior of the agriculture building. Write an introductory statement; include proper terminology.

2. The motion is debated by two or more members. Write down the discussion for and against the motion.

3. Lay the amended motion on the table. Write an introductory statement; include proper terminology.

4. A motion is made to have the annual parent–member banquet during National FFA Week in February. Write an introductory statement; include proper terminology.

5. A member moves to refer the motion to a committee. Write an introductory statement; include proper terminology.

6. A member discusses the motion to refer to a committee. Write discussion for the motion to refer.

7. A member moves the previous question. Write the proper terminology to make this motion.

8. A member moves to adjourn the meeting. Write the proper terminology for this motion.

## CONCLUSIONS

1. How can learning leadership in agriscience help you to become a better citizen in your community?

______________________________________________________________________

______________________________________________________________________

2. In what other meetings or organizations can you use parliamentary law?

______________________________________________________________________

______________________________________________________________________

3. List five individuals in your community whom you consider to be leaders. Give reasons for selecting each.

______________________________________________________________________

______________________________________________________________________

## OPTIONAL INTERNET SUPPLEMENT

The *Official FFA Manual* can be found at www.ffa.org. Search Resources. *Robert's Rules of Order* (www.rulesonline.com) provides rules of parliamentary procedure.

# SECTION 3

# Natural Resources Management

© Michelangelus/Shutterstock.com

# EXERCISE 9

## Watt's the Big Deal—Creating Electricity Out of Thin Air

### PURPOSE

The demand for electricity is at an all time high, and the price for it is also at its peak. Some farmers have begun to add wind energy to the list of commodities they produce on their land.

To create wind energy, large turbines are strategically placed in known windy areas. The wind catches the turbine's blades and creates lift causing them to rotate. The blades are connected to a drive shaft that turns an electrical generator.

In this lab, you will explore how electricity is created in an electrical generator.

### MATERIALS NEEDED

- 100 cm of copper wire
- 1 bar magnet
- 1 electrical meter
- 1 cardboard tube
- Duct tape
- Fan or turn table

### PROCEDURE

1. Wrap the copper wire around the cardboard tube about 20 to 25 times.
2. Connect both ends of the wire to the electrical meter (one end to each probe). The meter will register any electricity created. Movement on the meter will indicate electrical activity.
3. Hold the magnet and move it around the tube. Record the meter reading range.
4. Next take the magnet and move it through the tube. Try doing this at different speeds. Record the meter reading range.
5. Place the magnet on a flat surface and move the tube over it. Record the meter reading range.
6. Attach either the tube or the magnet onto a turn table or fan to find the position that creates the most electricity (optional). Record the meter reading range.

### OBSERVATIONS

| Procedure | Meter Reading Range |
|---|---|
| Procedure 3 | |
| Procedure 4 | |
| Procedure 5 | |
| Procedure 6 (opt.) | |

## CONCLUSIONS

1. Describe how you were moving both the tube and the magnet when the electrical meter was showing the most electrical activity (needle movement or highest readings).

2. When did the meter register the most electrical activity? Was it when the magnet was being moved around it or through it?

3. If a turn table or fan was used, did the extra speed create more energy? Explain why.

4. Using the information from question two, draw a sketch of a wind turbine that would produce the most electricity.

Source: USA. National Renewable Energy Laboratory. US Department of Energy. *I Renewable Energy And Energy Efficiency Science Projects*. By Cynthia Howell, Dr., Matt Kuhn, and Lung Linda. 2nd ed. Golden: n.p., 2007. *I Renewable Energy And Energy Efficiency Science Projects*. National Renewable Energy Laboratory. Web. 5 July 2013.

## OPTIONAL INTERNET SUPPLEMENT

The US Department of Energy's Energy Efficiency & Renewable Energy's Wind Program Web site includes a section that explains the process used to create electricity from wind (http://www1.eere.energy.gov/wind/).

Discovery Education (http://scienceofeverydaylife.discoveryeducation.com/) offers a Web site that allows students to build and test a virtual wind turbine. Select "Explore Virtual Labs" from the main menu to access 3M's Wind Energy Virtual Lab, or conduct an Internet search using the keywords "3M wind energy virtual lab" to locate the Web site.

Courtesy of DeVere Burton

# EXERCISE 10

## Getting Down and Dirty—Preventing Soil Erosion

### PURPOSE

This lab allows you to compare the rates and amounts of erosion that result from different land uses.

### MATERIALS NEEDED

- Six plant trays (or shoe boxes cut to 5 inches deep)
- Plastic sheeting to line trays
- Six sprinkling cans
- Six beakers
- Trowel
- Scissors
- Stopwatches
- Soil
- Sod (grass-covered soil from a lawn)
- Water
- Supports to form inclines for trays
- Measuring cup (mL)

### PROCEDURE

**1.** Preparing the materials

a. Cut a "V" notch in one end of each plant tray (or shoe box).

b. Line each tray with plastic, leaving an overhang of about 2 inches at the notch. This will serve as a spout.

c. Number and label each tray as you fill each of the six trays differently as follows:

1. moist soil packed firmly (bare soil)
2. sod (cover crop)
3. moist soil packed firmly; make packed furrows running across the tray's width (contouring)
4. moist soil packed firmly; make packed furrows running down the length of the tray (plowing up and down a hill)
5. moist soil packed firmly; form steps across the width of the tray with the trowel (terracing)
6. alternate three strips of moist packed soil with three strips of sod (strip cropping)

d. Position the trays on supports so the trays are tilted at an equal incline. Place beakers under the plastic overhangs at the "V" notches (spouts).

The following illustration offers a visual example of how to set up this experiment.

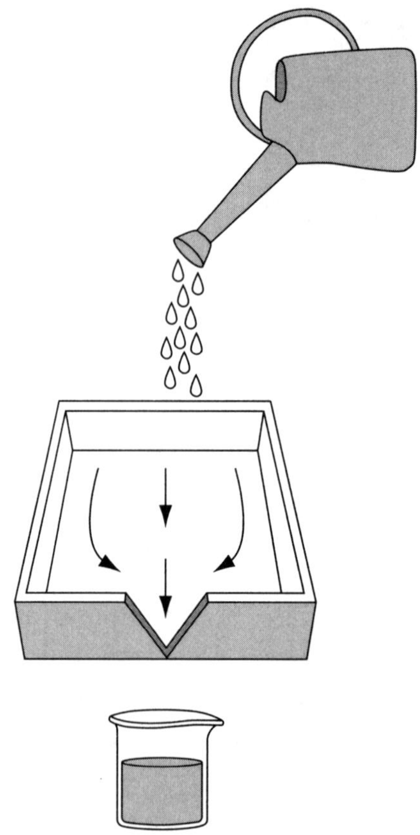



**2.** Conducting the experiment

a. Measure an equal amount of water into each sprinkling can.
b. With the help of fellow students, hold the sprinkling cans about 1 foot over the high point of each tray. Pour water steadily for 5 seconds onto each tray simultaneously.
c. In Observation 1, record the amount of time that water flows from the spout of each box.
d. After the water finishes draining, record the amount of runoff in each beaker in Observation 2.
e. Let the runoff settle, and then measure the volume of sediment in each beaker. Record the results in Observation 3.
f. Arrange the trays to increase the slope and repeat procedures 2a through 2e, recording the results in Observations 4 through 6.

## OBSERVATIONS

1.

| | Tray 1 | Tray 2 | Tray 3 | Tray 4 | Tray 5 | Tray 6 |
|---|---|---|---|---|---|---|
| Amount of Time Water Flows from the Spout | | | | | | |

2.

| | Beaker 1 | Beaker 2 | Beaker 3 | Beaker 4 | Beaker 5 | Beaker 6 |
|---|---|---|---|---|---|---|
| Amount of Runoff (mL) | | | | | | |

3.

| | Beaker 1 | Beaker 2 | Beaker 3 | Beaker 4 | Beaker 5 | Beaker 6 |
|---|---|---|---|---|---|---|
| Volume of Sediment (mL) | | | | | | |

4.

| | Tray 1 | Tray 2 | Tray 3 | Tray 4 | Tray 5 | Tray 6 |
|---|---|---|---|---|---|---|
| Amount of Time Water Flows from the Spout at Increased Slope | | | | | | |

5.

| | Beaker 1 | Beaker 2 | Beaker 3 | Beaker 4 | Beaker 5 | Beaker 6 |
|---|---|---|---|---|---|---|
| Amount of Runoff at Increased Slope | | | | | | |

6.

| | Beaker 1 | Beaker 2 | Beaker 3 | Beaker 4 | Beaker 5 | Beaker 6 |
|---|---|---|---|---|---|---|
| Volume of Sediment at Increased Slope | | | | | | |

## CONCLUSIONS

1. Which tray had the most runoff? Which tray lost the most soil?

2. Which tray's content best prevented erosion?

3. Is there a relationship between the time it took for each tray to finish draining, and how much soil washed off? Explain your answer.

4. How does the slope affect the soil loss?

5. How would you apply what you have learned from this experiment to soil conservation?

## OPTIONAL INTERNET SUPPLEMENT

Brighthub.com includes articles on environmental science topics including ways to reduce soil erosion. Search "preventing soil erosion" from the home page or navigate to the section on Environment.

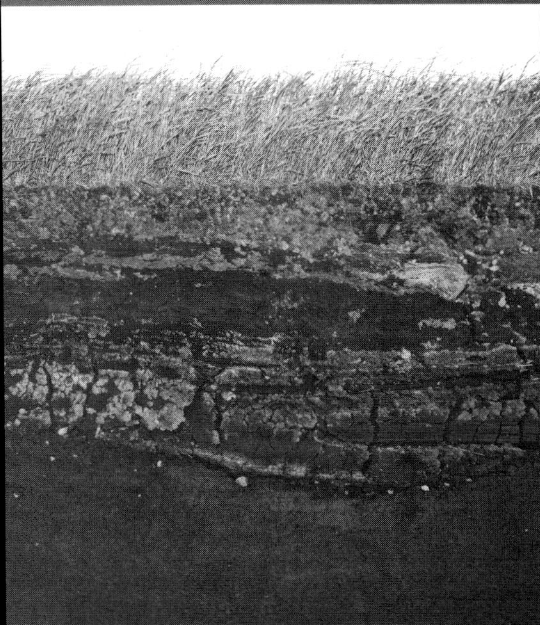



# EXERCISE 11

## Talking Dirty—Soil Fertility

### PURPOSE

This experiment will lead you to understand the value of soil nutrients by observing the effects of a nutrient deficiency in plants.

### MATERIALS NEEDED

- Seven identical 6-inch flower pots with drainage outlets at the bottom
- Macronutrient kit
- Pot chip or screen
- Vermiculite or perlite
- Mixing bin or box
- Pea and tomato seeds
- Nitrogen source—$NH_4NO_3$
- Phosphorus source—superphosphate
- Potassium source—KCl
- Calcium source—$CaCO_3$
- Materials for labeling pots
- Laboratory balance or scale
- Ruler or yardstick
- Personal Safety Set (includes rubber gloves and safety goggles or glasses)

## PROCEDURE

1. Place a screen or pot chip over the hole in the bottom of each pot.
2. Fill each pot with soil to 1 inch from the top.
3. Label the pots to correspond with the names of the treatments in column one of the following chart. One pot should be labeled for each treatment.

| Treatment | Nitrogen | Phosphorus | Potassium | Calcium |
|---|---|---|---|---|
| No Fertilizer | None | None | None | None |
| Complete Fertilizer | 1.1 g $NH_4NO_3$ | 1.25 g Superphosphate | .3 g KCl | None |
| Complete Fertilizer with Calcium | 1.1 g $NH_4NO_3$ | 1.25 g Superphosphate | .3 g KCl | 1.5 g $CaCO_3$ |
| Nitrogen Deficient | None | 1.25 g Superphosphate | .3 g KCl | None |
| Phosphorus Deficient | 1.1 g $NH_4NO_3$ | None | .3 g KCl | None |
| Potassium Deficient | 1.1 g $NH_4NO_3$ | 1.25 g Superphosphate | None | None |
| Calcium Only | None | None | None | 1.5 g $CaCO_3$ |



4. Weigh the amounts of fertilizer specified in columns 2, 3, 4, and 5 of the preceding chart. Dump the soil into a bin or

**SAFETY NOTE**

Wear eye protection glasses when handling any chemical solution. Handle chemical reagents with care, and follow your instructor's directions. Avoid contact with skin or eyes.

box and mix thoroughly with the fertilizer before returning the soil to the pot. Be sure not to mix materials from separate pots. Wash your hands after handling fertilizer.

5. Divide each pot into half with an imaginary line. Plant five pea seeds in one half and five tomato seeds in the other half. Water them in.
6. Place the pots in a location with adequate light and temperature for growing. Ensure that all pots have the same growing conditions.
7. Check the plants at regular intervals and water them as needed.
8. When the seeds have germinated and the plants are established, thin the plants so that each pot has the same number of each kind of plant.
9. Observe the plants each week, making note of differences in color, height, width, and any other differences between pots. Record this information in the table in Observation 1.
10. At the conclusion of the experiment, harvest all plants by pulling them up by the roots. Weigh all the plants of the same kind in each pot together and record the green weights on the chart in Observation 2.
11. Let the plants dry in the sun for 4 days. Be sure to keep the plants from different pots separated. Weigh each kind of plant in each pot again separately. Record the dry weights on the chart in Observation 2.

## OBSERVATIONS

**1.** In the following table, record your observations of rate of growth, color, and so on.

| | Treatment | | | | | | |
|---|---|---|---|---|---|---|---|
| **Observations** | **No Fertilizer** | **Complete Fertilizer** | **Complete Fertilizer with Calcium** | **Nitrogen Deficient** | **Phosphorus Deficient** | **Potassium Deficient** | **Calcium Only** |
| Week 1 | | | | | | | |
| Week 2 | | | | | | | |
| Week 3 | | | | | | | |
| Week 4 | | | | | | | |
| Week 5 | | | | | | | |
| Week 6 | | | | | | | |
| Week 7 | | | | | | | |
| Week 8 | | | | | | | |

**2.** In the following chart, record the harvest weight for each kind of plant for each treatment.

| Treatment | Green Weight | | Dry Weight | |
|---|---|---|---|---|
| | Peas | Tomatoes | Peas | Tomatoes |
| No Fertilizer | | | | |
| Complete Fertilizer | | | | |
| Complete Fertilizer with Calcium | | | | |
| Phosphorus Deficient | | | | |
| Potassium Deficient | | | | |
| Calcium Only | | | | |



## CONCLUSIONS

**1.** What were the signs of deficiencies in the plants you grew, if any?

a. Nitrogen ______________________________

______________________________

b. Phosphorus ______________________________

______________________________

c. Potassium ______________________________

______________________________

d. Calcium ______________________________

______________________________

**2.** Which plants seemed to grow best in the soil to which calcium was added?

______________________________

**3.** How did the green weights compare to the dry weights?

______________________________

______________________________

______________________________

## OPTIONAL INTERNET SUPPLEMENT

Iowa State University's Department of Agronomy Web page on Soil Fertility provides a good resource for learning about soils. Go to www.agronext.iastate.edu/soilfertility/ Search "Do You Know The Story of Soil?"

© mackzhappy/Shutterstock.com

# EXERCISE 12

## Hold the Soil Please—Growing Plants without Soil

### PURPOSE

In this exercise you will demonstrate the art of growing plants without soil (hydroponics).

### MATERIALS NEEDED

- Hydroponics Experiment Kit
- Personal Safety Set

**SAFETY NOTE**

Wear eye protection glasses when handling any chemical solution. Handle chemical reagents with care, and follow your instructor's directions. Avoid contact with skin or eyes.

### PROCEDURE

1. Fill the transparent plastic culture chamber with deionized water to the ridge located near the top edge.
2. Add 50 drops of each of the following nutrient solutions to the water-filled chamber and stir with a clean glass rod.
   a. $Ca(NO_3)_2$ solution (calcium nitrate)
   b. $KNO_3$ solution (potassium nitrate)
   c. $MgSO_4$ and $KH_2PO_4$ solution (magnesium sulphate and potassium phosphate)
3. Add 5 drops of the iron chelate solution to the chamber.
4. Add 5 drops of the trace elements solution to the chamber.
5. Carefully place a seed inside the slit of a germinating plug. This should then be inserted into a hole provided in the white germinating float. When each of the 12 holes has been filled with seeds (and plugs), carefully place the float into the chamber so that all plugs are in contact with the liquid in the chamber.
6. Replace the opaquing shield on the lower portion of the culture chamber, if it has been removed. The shield will inhibit the growth of algae in the solution and should be removed to observe root growth.
7. Replace the vented cover and allow the plants to germinate. A south or west window is the best location for the culture chamber.
8. As the solution is used by the plants, it should be brought back to the original level by adding deionized water. After the seeds have germinated, replace the nutrients and trace elements once a week.
9. Observe the growth of the plants daily for 2 weeks and record the progress on the chart in Observation 1.

## OBSERVATIONS

1. Record the daily progress of the plants in the following chart.

| Day | Plant Growth Conditions | | | |
|---|---|---|---|---|
| | **Length of Shoots and Roots** | **Coloration of Roots** | **Coloration of Leaves** | **Other Conditions** |
| 1 | | | | |
| 2 | | | | |
| 3 | | | | |
| 4 | | | | |
| 5 | | | | |
| 6 | | | | |
| 7 | | | | |
| 8 | | | | |
| 9 | | | | |
| 10 | | | | |
| 11 | | | | |
| 12 | | | | |
| 13 | | | | |
| 14 | | | | |

## CONCLUSIONS

1. How can the plants germinate without soil?

2. Did you observe any deficiencies in the plants?

3. What color were the roots of the plants?

4. What possibilities do you see for the use of hydroponics in the future?

## OPTIONAL INTERNET SUPPLEMENT

University of Virginia (http://www.pubs.ext.vt.edu): Search-Home Hydroponics. This Web site shows the different type of home hydroponic systems.

© Rob Hainer/Shutterstock.com

# EXERCISE 13

## There's Something Fishy Here—Fish Anatomy

### PURPOSE

Exercise 13 will let you become familiar with the major internal and external features, structures, and functions of fresh water fish.

### MATERIALS NEEDED

- Dissection Lab Perch Kit
- Dissection set
- Magnifying glass
- Personal Safety Set

**SAFETY NOTE**

Always wear eye protection glasses and rubber gloves when handling dissection specimens. Wash hands before leaving the laboratory.

### PROCEDURE

**1.** Study and become familiar with the following terms:

a. **dorsal**—back or upper
b. **ventral**—abdominal or lower
c. **anterior**—forward or front
d. **posterior**—hind or rear
e. **median**—toward the middle

**2.** Refer to Figure 12-7 in Unit 7 of your textbook to study the parts of fresh water fish and learn their functions as defined.

**External Anatomy**

a. **mouth**—at anterior end of the animal. The position of the mouth reflects the manner whereby the fish feeds. The fish possessing a terminal mouth feeds by overtaking prey while swimming. The fish possessing a dorsal mouth is usually a surface feeder. If the mouth is ventral, the fish is usually a bottom feeder.
b. **nostril**—two pairs, each pair opens into an olfactory (smell) chamber and allows water to pass through the chamber. The movement of water is essential to odor perception.
c. **eye**—two large spherical eyes are located on each side of the head.
d. **gill openings**—allow water passing through the gills to exit.
e. **operculum**—a semicircular flap that covers the gills.
f. **pectoral fins**—paired fins found on either side of the perch.
g. **pelvic fins**—paired fins found close together on the ventral surface of the fish trunk.

h. **anterior dorsal fin**—single fin found on the dorsal side of the fish and supported by 13 to 15 spiny rays.
i. **posterior dorsal fin**—single fin located directly behind the anterior dorsal fin and supported by 7 to 8 soft rays.
j. **caudal fin**—the fleshy end of the tail is surrounded by this fin; the type of caudal fin found on the perch is called *homocercal.*
k. **anal fin**—single fin located behind the anus on the ventral side and supported by soft rays.
l. **anus**—opening through which feces are eliminated from the fish body.
m. **skin**—envelope for the body and the first line of defense against disease; it also affords protection from, and adjustments to, environmental factors that influence life.
n. **mucous glands**—glands located in the skin, the secretions of which give the slimy touch and odor to the fish.
o. **fish scales**—thin, flexible plates that provide a flexible "armor" to protect the fish. They are used in growth and age studies. The scales of perch are called *ctenoid* because the exposed portion bears tiny, spiny structures called *cteni* (teeth).
p. **lateral line system**—a sensory organ system restricted to fish and aquatic stages of amphibians; it enables a fish to detect localized disturbances from currents or vibrations caused by fixed or moving objects.

**3.** Study external anatomy
a. Place the perch on the dissection tray, side down, with the head facing to the left, dorsal fins up.
b. Locate and identify the external features of the perch. Label these on the diagram in Observation 1.
c. Using forceps, carefully remove a fish scale from the trunk region and examine it under a magnifying glass. Identify the cteni. Complete Observation 2.

**4.** Study the internal parts of the perch. Each part's function is discussed in the appropriate section under "Internal anatomy."

**Internal Anatomy**

a. **muscular system**—controls body movements.
   1. With a scalpel, carefully cut a large (2-inch) square window on the lateral surface of the perch directly below the anterior dorsal fin.
   2. Using forceps, carefully peel away the skin to expose the muscle layer beneath. Identify and label the trunk muscles on the diagram in Observation 3.

Trunk muscles are attached to the skeleton and are of primary use in producing body movement.

b. **respiratory system**—serves in the exchange of gases with the environment.
   1. Using scissors, carefully cut the operculum at its base on the head to expose the gills. Observe the figure in the kit.
   2. Extend the cut posteriorly from the base of the angle of the jaw around the gills. Raise and remove the operculum.
   3. Pry the mouth open with a probe and observe the mouth and pharynx (pharyngeal cavity). Feeding is accomplished by sucking prey into the mouth, unlike fish that bite their prey.
   4. Identify the following parts of the respiratory system on the fish and label them on the diagram in Observation 3.
      a. **palate**—solid roof extending from the mouth cavity to the pharynx.
      b. **gills**—organs of respiration with thin walls that are supplied with blood capillaries, as are our lungs.
      c. **gill arches**—serve as attachments for the gills.
      d. **gill filaments**—tiny fingers running at right angles to the length of each gill. Oxygen dissolved in water diffuses through the thin wall of each gill filament into capillaries, and carbon dioxide diffuses in the opposite direction.
      e. **gill rakers**—bony structures located on each gill arch and pointing toward the pharynx. Collectively, they act as a strainer preventing food and other particles from being carried across the gills, which could cause injury to the gills.

c. **circulatory system**—the transport system of the body supplies oxygen and nourishment to body cells and removes metabolic wastes.
   1. Remove the exposed pectoral, pelvis, and anal fins.
   2. Using scissors, cut through the lateral side of the body wall.
   3. Remove the section of body wall cut.
   4. Hold the perch in place by pinning the head, tail, and fins to the dissection tray.
   5. Identify the following parts of the circulatory system on the fish and label them on the diagram in Observation 3.
      a. **heart**—a two-chambered, muscular organ that receives blood from the body, circulates it to the gills for oxygenation to the body, and then back to the heart. Blood in the fish passes from the atrium to the ventricle without being oxygenated.

b. **arteries**—carry blood from the heart to the body tissues.
c. **veins**—carry blood to the heart from the body tissues.
d. **spleen**—a related circulatory organ; an elongated structure lying on the posterior dorsal surface of the stomach. It is concerned with the production and maintenance of blood cells.

d. **digestive system**—functions to both mechanically and chemically break down food into simple components easily absorbed by the blood and passed throughout the body to provide body cells with nourishment.

1. Identify the following parts of the digestive system on the fish and label them on the diagram in Observation 3.
   a. **esophagus**—a short, straight tube leading from the mouth to the stomach.
   b. **stomach**—receives food from the esophagus and digests the food's protein components.
   c. **intestinal diverticula**—three blind sacs located at the anterior-most part of the duodenum, where it connects to the stomach. The pancreatic and bile ducts open here.
   d. **duodenum**—partially digested food is passed from the stomach to this organ for final digestion. The duodenum forms an S-shaped loop from its origin at the stomach.
   e. **intestine**—final digestion and absorption of food from the duodenum takes place here.
   f. **fat bodies**—white masses located around the intestine; they function to store food reserves.
   g. **anus**—outside opening of the intestine through which solid waste is passed.
   h. **digestive glands**
      1. **liver**—reddish-brown, lobed organ situated just anterior and dorsal to the stomach. It produces bile, that aids in fat digestion and the regulation of nutrient levels in the blood.
      2. **gallbladder**—greenish sac between the lobes of the liver; stores bile before release to the duodenum.
      3. **pancreas**—a small, folded organ located at the ventral border of the duodenum; manufactures pancreatic juices that further digest proteins.

e. **excretory system**—removes metabolic waste products from the blood.
  1. Identify the following parts of the excretory system on the fish and label them on the diagram in Observation 3.
     a. **kidneys**—paired, elongated masses pressed against the dorsal body wall.
     b. **collecting ducts**—tubes that drain each kidney. In the male, they connect to the urinary bladder. In the female, they connect the kidney to a -urogenital sinus where the urinary bladder is incorporated into the oviduct.
     c. **urinary papilla**—external projection of the male urinary bladder. Waste products (urine) are eliminated from the body through a urinary pore. In the female, urine is eliminated through the abdominal pore.
     d. **urinary bladder**—a small sac of the most posterior, ventral area of the body cavity; functions to store urine.

f. **reproductive system**—responsible for the production of sex cells and their delivery to produce offspring; sex organs are separate in the perch.
  1. Identify the following parts of the reproductive system on the fish, and label them on the diagram in Observation 3.
     a. **testes** (male)—paired, lobulated structures, posterior to the duodenum and stomach, and dorsal to the intestine. They produce sperm, that is transported to the outside by a duct called the *vas deferens* and eliminated through the genital pore.
     b. **ovary** ( female)—single, large sac located in the same position in the body as are the testes in the male. The ovary produces white spherical eggs, that fill the cavity to capacity in the spring.
     c. **oviduct**—a funnel-like structure where the urinary bladder is incorporated. Eggs are expelled through the abdominal pore.
     d. **other parts**
        1. **gas bladder** or **swim bladder**—helps regulate depth independently of body motion; located ventral to the kidney and dorsal to the sex organs.

# OBSERVATIONS

**1.** Label the external parts of the perch on the following diagram.

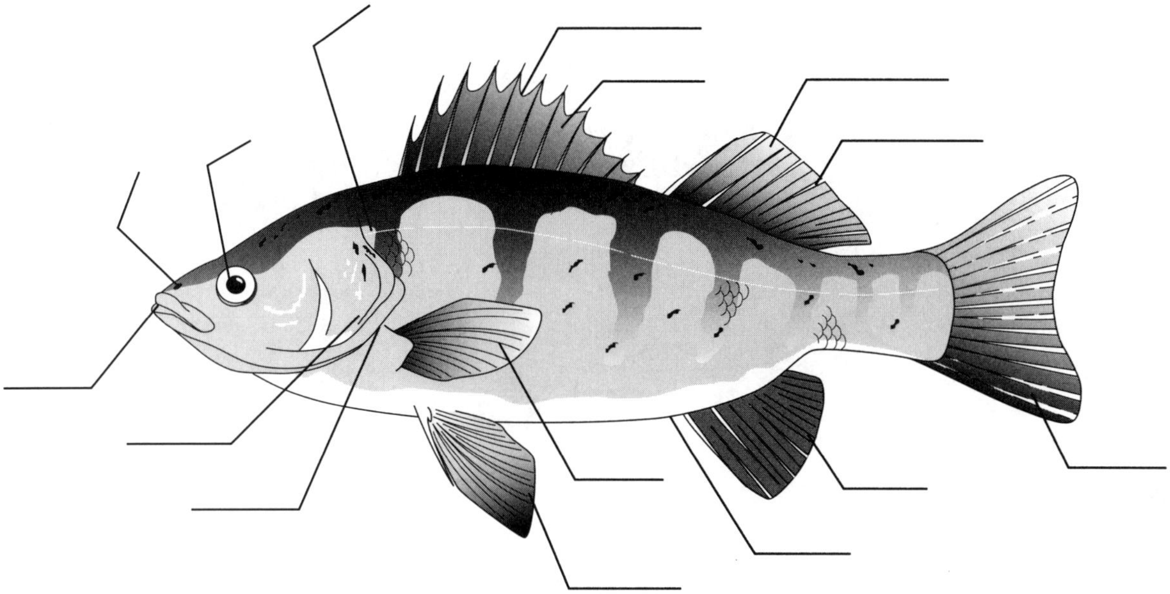


**2.** Draw a diagram of the fish's scales. Be sure to show how the scales overlap.

**3.** Label the internal parts of the perch on the following diagram.

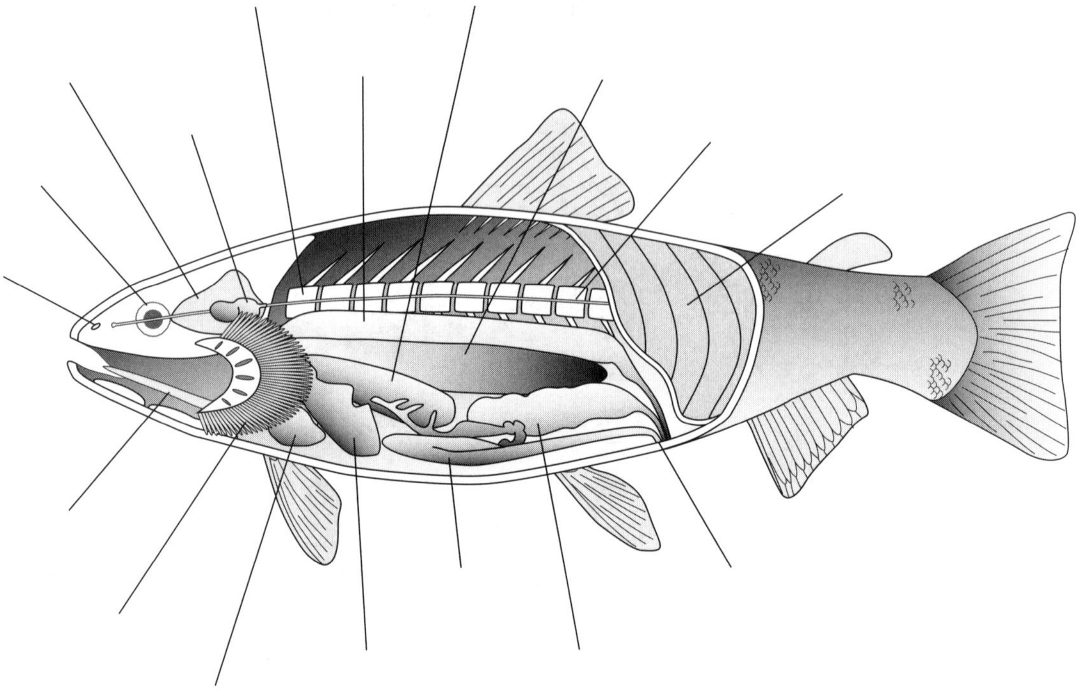



**4.** Is your fish male or female? How can you tell?

______________________________________________

______________________________________________

______________________________________________

**5.** Cut open the stomach of the fish. What types of food do you find there?

______________________________________________

______________________________________________

______________________________________________

## CONCLUSIONS

1. How many fins does your fish have? What purpose is served by the fins?

2. How do the nostrils aid the fish in smelling?

3. Based on the mouth location, what type of feeder is the perch?

4. How is the heart of a fish different from yours?

5. What is the function of the gas bladder?

## OPTIONAL INTERNET SUPPLEMENT

Oracle® ThinkQuest Education Foundation's Web site includes a virtual fish dissection activity called Salmonids: In Troubled Waters. Navigate to library.thinkquest.org and search "salmonid dissection."

# SECTION 4

# Integrated Pest Management

Courtesy of DeVere Burton

# EXERCISE 14

## Are You a Competitor? Weed Identification Lab

### PURPOSE

Believe it or not, you are in competition with pests every day. Anyone who ever weeded a garden understands they are seriously out-numbered. Fortunately for us, integrated pest management strategies can help. The first step in dealing with pests is to know your enemy. This is the most difficult and most important part of the process. In this lab, you will be gathering common weeds found in your community and then identifying them by their common and scientific names.

### MATERIALS NEEDED

- A large amount of newspaper
- Two pieces of tempered hard board, peg board, or plywood (~12"×18" rectangles).
- Cardboard cut into ~12"×18" rectangles
- Tie-downs or bungee cords for strapping the weed portfolio tightly.
- Large pieces of construction paper for mounting the specimens.

### PROCEDURE

Allow several days for this lab.

1. Collect 20 weed specimens from near your home. This should be done as homework. Use a shovel to get all of the plant (roots, stem, leaves, flowers, fruit and seeds). Not all plants will have the last three items. Place the weed into a plastic sack. Leave the soil on the root ball until you are ready to press it, and then remove it.
2. While the specimen is still fresh identify it. It is much easier to identify a fresh plant than it is a pressed and dried one. This can be done using the Internet (key words), weed identification guide, or a weed identification manual.
3. To mount the weed specimen, place the plant between several pages of newspaper.
   - Large plants can be bent a couple of times so it will fit on the page.
   - Show the underside of some of the leaves so that both sides of the leaves can be seen.
   - Remove as much soil as possible from the roots. Spread the roots out as much as possible.
   - Seed samples collected should be placed in a zip style bag and saved for mounting later.
4. Now that the plant is inside three or four layers of news paper, it will be placed between two pieces of cardboard. All of your samples should be layered this way and placed in-between the outer cover (constructed of lathe or thin plywood). Next, use the tie-downs or bungee cords to wrap the specimens inside the two pieces of wood. Finally place

- Plastic bags for collecting (old grocery bags work well and keep the specimens moist)
- Shovel
- Adhesive for mounting the specimens
- Blank adhesive labels
- 20 weeds collected near your home
- Internet access
- Weed Identification manuals
- Small zip style bags

a heavy object on top of your sample portfolio for a few days.

- Change the newspaper and cardboard as often as needed (at least once a day at first). Extremely green plants are hard to preserve.

**5.** Once the weeds are dry, carefully attach them to a large sheet of construction paper. Take care when doing this because, the dried plants are very brittle. Place a plant identification label on the back of the page.

- The label should include; common names, Genus and species, date, and location of collection.
- Be sure to include the weed's seeds and fruit on the mounting page.
- Your final portfolio should include 20 mounted plants with labels. The pages should be numbered, and they should correspond to a table of contents.

***Note:*** In this lab, your observations and conclusions are included in the overall process.

## OPTIONAL INTERNET SUPPLEMENT

University of California Davis Integrated Pest Management Online (www.ipm.ucdavis.edu) is an online weed identification site. USDA Natural Resources Conservation Service (NRCS) has a Plants Database (plants.usda.gov) that lists nocuous weeds by state.

Boise National Forest

# EXERCISE 15

## Don't Bug Me—Insect Anatomy

### PURPOSE

This exercise enables you to recognize and identify the major external and internal features and structures of the grasshopper and explain their functions.

### MATERIALS NEEDED

- Dissection Manual Grasshopper Kit
- One pair of fine dissection scissors
- One sharp probe
- Dissecting microscope
- Personal Safety Set
- Internet access
- Text book

**SAFETY NOTE**

Always wear eye protection glasses and rubber gloves when handling dissection specimens. Wash hands before leaving the laboratory.

### PROCEDURE

1. Study the information regarding anatomical orientation terminology and the biology of the grasshopper outlined in the grasshopper kit.
2. Study the information regarding body shape and the exoskeleton outlined in the grasshopper kit.
3. External anatomy
    a. Place the grasshopper on the dissection tray so its head is to your left and its legs are down.
    b. Locate and identify the external features of the grasshopper by referring to the diagrams provided in the grasshopper kit.
    c. Use the dissection microscope to observe some of the smaller details.
    d. Using the text or an insect anatomy Web site, locate and label the external parts on the diagram in Observation 1.
4. Identify the sex of your grasshopper using the external anatomical features detailed in the kit. Record in Observation 2.
5. Internal anatomy
    a. Snip off the legs, wings, and antennae of the grasshopper at their bases.
    b. Using pointed iris scissors, insert one point of the scissors on the left side in the intersegmental membrane between the 10th and 11th abdominal segments just below the cercus. Cut anteriorly up the left side of the grasshopper at a level of approximately 1/8 inch above the level of the spiracles.
    c. Continue to cut at the same approximate level all the way to the head. (Observe the diagram in the kit.)

d. Make a similar cut at the same approximate level on the right side of the insect, extending the cut anteriorly from the 10th and 11th abdominal segment to the head. Join both cuts so the grasshopper is completely bisected laterally.
e. Carefully lift the posterior part of the dorsal body dissection, working it toward the head, free from the larger ventral body dissection. No difficulties will be encountered until you reach the thorax. Here, cut through the dorsoventral and oblique body muscles and ligaments. In the head, free the head muscles as close to the dorsal dissection body wall as possible.
f. Remove the dorsal body dissection and place both the dorsal and ventral body dissections in the dissection tray, using insect pins to securely position both dissections.
g. Orient both dissections (ventral and dorsal) in the dissection tray with the head region facing away from you.
h. Locate and identify the internal structures of the grasshopper by referring to the diagrams provided in the grasshopper kit. Study the various systems as discussed in the kit.
i. Using an insect anatomy Web site, label the various internal parts on the diagrams in Observation 3.

## OBSERVATIONS

**1.** Label the external parts of the grasshopper on the following diagram.

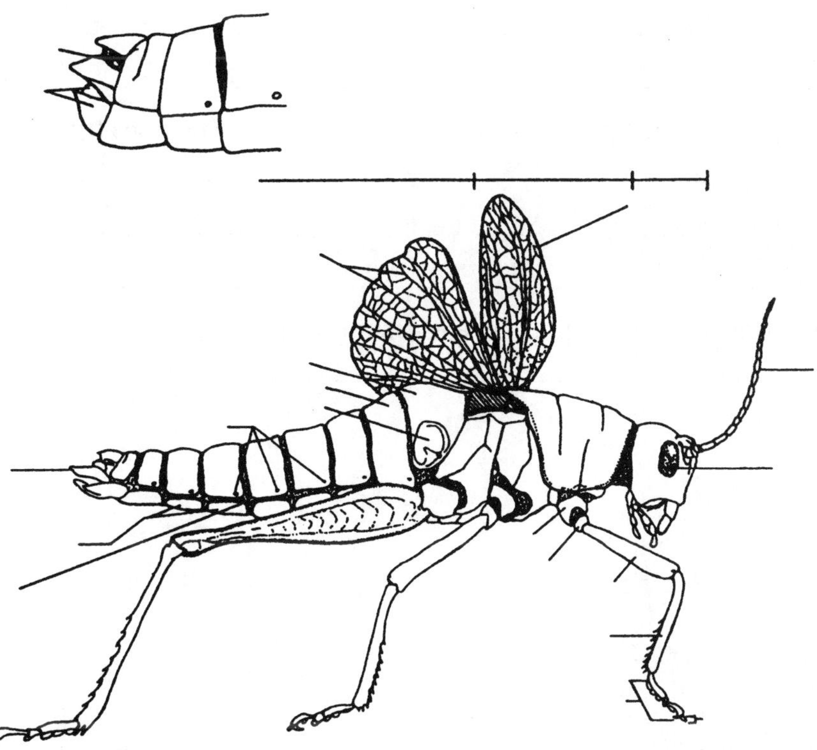

2. Circle the sex of your grasshopper. Male/Female

3. Label the internal parts of the grasshopper on the following diagrams:

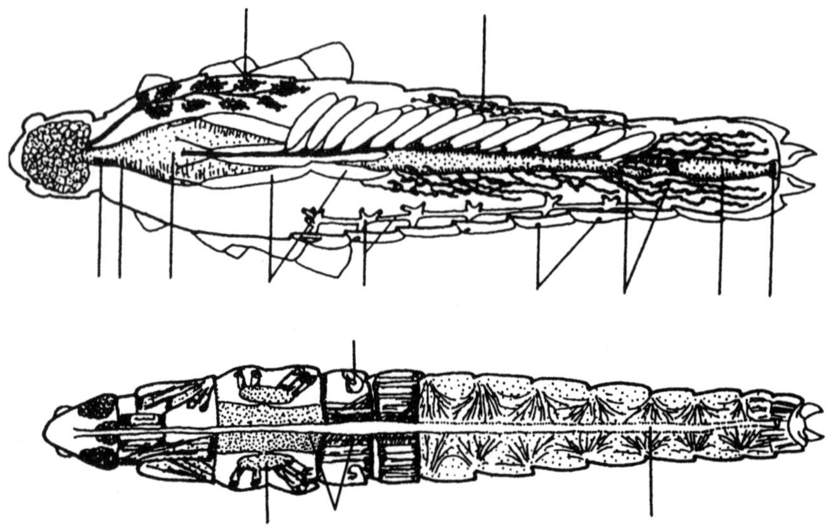

## CONCLUSIONS

1. How is the exoskeleton of a grasshopper different from the skeleton of a human? How are they alike?

_______________________________________________

_______________________________________________

_______________________________________________

2. What three types of mouthparts may insects have? What type does the grasshopper have?

_______________________________________________

_______________________________________________

_______________________________________________

3. What are the three main body parts of a grasshopper? _______________________________________________

_______________________________________________

**4.** Using the Internet or other sources provided by your instructor, explain the functions of the following:

a. antenna ____________________
b. mandibles ____________________
c. pretarsus ____________________
d. spiracle ____________________
e. ovipositor ____________________
f. sclerites ____________________
g. crop ____________________
h. malpighian tubules ____________________
i. ocellus ____________________
j. cercus ____________________

## OPTIONAL INTERNET SUPPLEMENT

U.S. Environmental Protection Agency (EPA), Pesticides: Topical & Chemical Fact Sheets and the EPA's integrated pest management and best practices guidelines can be found at this Web site. (http://www.epa.gov/pesticides/factsheets/ipm.htm)

Insectidentification.org provides an insect identification key.

CDC/ Dr. Gary Alpert - Urban Pests - Integrated Pest Management (IPM)

# EXERCISE 16

## Stuck on You—External Parasites

### PURPOSE

External Parasites can cause discomfort and disease in domestic animals. Learning to remove these parasites allows you to better care for your animals.

### MATERIALS NEEDED

- Compound microscope (used for most lice and fleas)
- Dissecting scope (sufficient for larger lice and engorged ticks)
- Insecticide powder
- Animal comb or any inexpensive plastic hair comb
- Smooth dark material (paper or plastic)
- 70 percent ethyl alcohol or isopropyl (rubbing) alcohol
- Glass jar
- Water or oil (baby oil)
- Microscope slides and cover slips
- Dissecting needle
- Personal Safety Set
- Internet access (optional)

**1.** In general, lice, fleas, and ticks cause
   a. ______________________________
   b. ______________________________
   c. ______________________________

**2.** What diseases can fleas transmit to other animals and humans?
   a. ______________________________
   b. ______________________________

**3.** What diseases are ticks responsible for transmitting?
______________________________

### PROCEDURE

***Note:*** Because of exposure issues, your instructor or a licensed animal health care professional will perform the parasite removal procedure (steps 1–4). Ticks may harbor a life threatening illness. Gloves must be worn by all individuals handling a parasite.

**1.** Dust the animal with an appropriate insecticide powder that will kill or immobilize the parasites. The insecticide used must be approved and safe for the breed, sex, age, and function of the animal. Carefully follow the directions on the label to minimize any harmful effects to the animal.

**2.** Comb the parasites out of the coat of the animal onto a dark, smooth surface.

**3.** Bunch the parasites together and transfer to a glass jar containing alcohol.

**4.** Slowly pull away the tick to avoid leaving the mouthparts in the skin or the animal. Place the removed tick in a glass

**SAFETY NOTE**

When dealing with parasites and blood exposure, always wear medical gloves and eye protection glasses, wash hands and any area that has been exposed with an antibacterial soap.

jar containing alcohol. A dissecting scope can be used to identify the engorged nymph or adult tick.

5. To examine under a microscope, remove a few specimens from the alcohol with finely pointed forceps and place in a drop of water or oil on the microscope slide. Place a cover slip on top, being careful not to push down too hard. Morphological details of most species of lice and fleas can be seen with low magnification of a compound light microscope. A dissecting scope must be used for larger lice. Nymph and adult ticks that are attached and engorged can be removed from the skin with forceps.
6. The identification of lice and fleas as to genus and species is difficult and requires an entomologist specialized in the subject. In this exercise, we are just distinguishing between lice, fleas, and ticks.

## OBSERVATIONS

Sketch your observations in the boxes below; include a drawing of a louse, a flea, and a tick. If any of these parasites were not found during the removal process, find an image of it on the internet.

## OPTIONAL INTERNET SUPPLEMENT

Texas Agricultural Extension Service (http://ector.agrilife.org)
Search-publications/animal health/managing external parasites
This Web site provides information about external parasite management.

# SECTION 5

## Plant Sciences

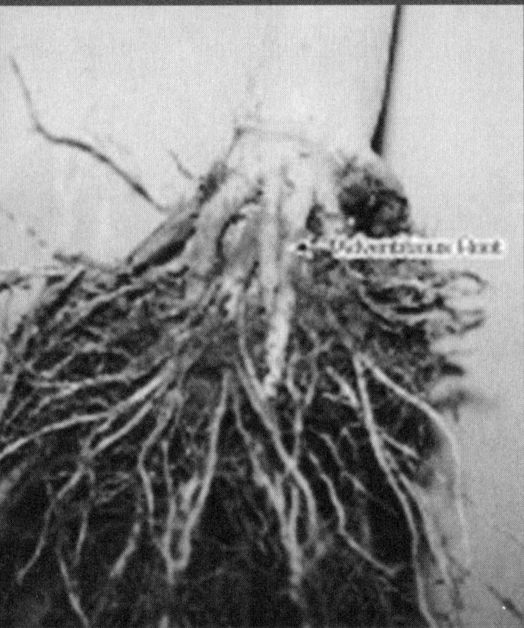

Courtesy of DeVere Burton

# EXERCISE 17

## Putting Down Some Roots—Root Anatomy

### PURPOSE

The purpose of this exercise is to become familiar with the structures and functions of monocot and dicot roots.

### MATERIALS NEEDED

- Monocot root slide
- Dicot root slide
- Compound microscope
- *Agriscience Fundamentals and Applications 6th* edition Unit 5
- Internet Access
- Set of colored pencils

### PROCEDURE

1. Study the functions of the monocot root parts described here, and identify them on the labeled diagram of a root cross-section. Refer to Figure 15-13 in your textbook that illustrates the cross-section of a stem, including the xylem, phloem, and vascular system.
   a. **epidermis**—the outer layer of a monocot root functions to protect the root from loss of water and from plant infections.
   b. **mechanical tissue (hypodermis)**—a layer of thick-walled cells; serves as support and protection for the root.
   c. **cortical parenchyma**—a thick layer of root cells lying between the hypodermis and the xylem.
   d. **xylem**—composed of tubes of varying size that carry water absorbed by the root hairs from the roots to the other parts of the plant.
   e. **phloem**—the small, closely-packed cells found between the xylem and the pith; they carry food produced by photosynthesis in the leaf to the roots. The phloem is composed of sieve tubes that are elongated, square-cornered cells of varying size and with perforated ends.
   f. **pith**—the region of root cell maturation; composed of undifferentiated parenchyma cells, that are large and not closely packed.
   g. **root tip**—the very short portion of the extreme free end of a root, where growth, water absorption, and the differentiation of the primary tissues take place.

   h. **root cap**—thimble-shaped mass of loose cells that protect the growing point of the root.
   i. **root hairs**—long, fingerlike structures projecting between the tiny particles of soil; absorb water for the entire plant.
2. Place the monocot root cross-section on the microscope stage centered over the light hole and fasten with the stage clips. Identify the root parts under the microscope.
3. Label the parts of the monocot root on the diagram in Observation 1 and color the parts as follows:
   a. epidermis—light green
   b. xylem—red
   c. phloem—yellow
   d. mechanical tissue—brown
   e. cortical parenchyma—orange
4. Study the functions of the dicot root parts described here, and identify them on the labeled diagram of a root cross-section.
   a. **epidermis**—the outer layer of a dicot root functions to protect the root from loss of water and plant infections.
   b. **parenchyma**—a thick layer of root cells lying between the epidermis and the endodermis.
   c. **phloem**—the outer layer of the stele ((the central core of vascular tissue in a plant stem) serves as a food transport system.
   d. **pambium**—cylinder of thin-walled cells that gives rise to new xylem to the inside and phloem to the outside. The second layer of the stele found between the phloem and the xylem.
   e. **xylem**—tissue that carries water from the roots to the other parts of the plant.
   f. **pericycle**—sheath of small, parenchymatous cells that surrounds the xylem and phloem.
5. Place the dicot root cross-section on the microscope stage centered over the light hole and fasten with the stage clips. Identify the root parts under the microscope.
6. Label the parts of the dicot root on the diagram in Observation 2, and color the parts as follows:
   a. epidermis—yellow
   b. parenchyma—orange
   c. endodermis—light green
   d. phloem—dark green
   e. cambium—red
   f. xylem—brown
   g. pericycle—blue

## OBSERVATIONS

1. Label the parts of the monocot root on the following diagram and color the parts as outlined in Procedure 3.

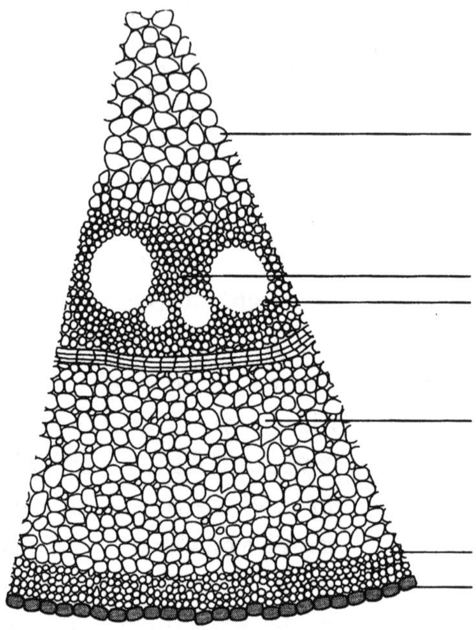



2. Label the parts of the dicot root on the following diagram and color the parts as outlined in Procedure 6.

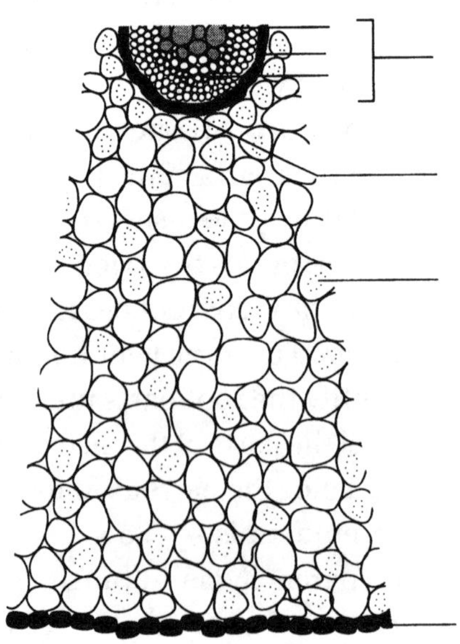

## CONCLUSIONS

**1.** What composes most of the cells in a monocot root and in a dicot root?

______________________________

______________________________

______________________________

**2.** What are the functions of the following in a plant root?

a. xylem ______________________________

______________________________

b. phloem ______________________________

______________________________

c. cambium ______________________________

______________________________

d. epidermis ______________________________

______________________________

e. root cap ______________________________

______________________________

f. root hairs ______________________________

______________________________

**3.** What functions do roots serve?

______________________________

______________________________

______________________________

______________________________

______________________________

______________________________

## OPTIONAL INTERNET SUPPLEMENT

University of Rhode Island (http://www.uri.edu/cels/bio/plant_anatomy/images.html): Search—Anatomy of Roots.

This Web site provides images of microscopic plant root cells.

© Madlen/Shutterstock.com

# EXERCISE 18

## It Stems from—Stem Anatomy

### PURPOSE

In this exercise you will become familiar with the structures and functions of monocot and dicot stems.

### MATERIALS NEEDED

- Monocot stem slide
- Dicot stem slide
- Compound microscope
- Set of colored pencils

### PROCEDURE

**1.** Study the functions of the monocot stem parts described here and identify them on the labeled diagram of a stem cross-section. Refer to Figure 15-8 in your textbook that illustrates the important parts of stems.

a. **epidermis**—the hard, green, outer layer of a monocot stem functions to protect the stem from loss of water and from plant infections.

b. **fibro vascular bundle**—collectively, fibrous strands found in the monocot stem and composed of xylem, phloem, and mechanical tissue.

c. **xylem**—the inner part of the vascular bundle; carries water and nutrients from the roots to the other parts of the plant.

d. **phloem**—the small cells on the outer part of the vascular bundle; they carry food produced by photosynthesis in the leaf to other parts of the plant. The phloem is -composed of sieve tubes that are elongated, square-cornered cells of varying sizes and with perforated ends.

e. **mechanical tissue**—a layer of thick-walled cells surrounding the vascular bundle; provides support and protection.

f. **parenchyma**—cells that make up the pith of a monocot plant.

g. **air space**—air spaces found in the pith of the monocot stem.

**2.** Place the monocot stem cross-section on the microscope stage centered over the light hole and fasten with the stage clips. Identify the stem parts under the microscope.

**3.** Label the parts of the monocot stem on the diagram in Observation 1 and color the parts as follows:
   a. epidermis—light green
   b. xylem—red
   c. phloem—yellow
   d. mechanical tissue—brown
   e. air space—white
   f. parenchyma—orange

**4.** Study the functions of the dicot stem parts described below and identify them on the labeled diagram of a stem cross-section.
   a. **cork**—outer layer of the bark; serves to protect the stem of the plant from disease and loss of water.
   b. **cork cambium**—inner layer of the bark; gives rise to cells that become impregnated with a waxy substance, forming a waterproof layer that becomes cork.
   c. **cortex**—cells that are found inside the cork cambium and serve as a food storage area.
   d. **phloem**—the layer of cells found inside the cortex and serves as a food transport system. The phloem consists of phloem rays and phloem fiber.
   e. **cambium**—cylinder of thin-walled cells that gives rise to new xylem to the inside and phloem to the outside.
   f. **xylem**—tissue that carries water and forms the wood of a tree. The growth rings of a tree are found here.
   g. **pith**—the center of a dicot stems; provides support to the plant during the first year of growth.
   h. **vascular rays (phloem rays** and **xylem rays)**—horizontal strips of thin-walled cells; provide for the transverse conduction of water and nutrients along the tissues of the stem.

**5.** Place the dicot stem cross-section on the microscope stage centered over the light hole and fasten with the stage clips. Identify the stem parts under the microscope.

**6.** Label the parts of the dicot stem on the diagram in Observation 2 and color the parts as follows:
   a. cork—yellow
   b. cork cambium—orange
   c. cortex—light green
   d. phloem—dark green
   e. cambium—red
   f. xylem—brown
   g. pith—blue

## OBSERVATIONS

1. Label the parts of the monocot stem on the following diagram and color the parts as outlined in Procedure 3.

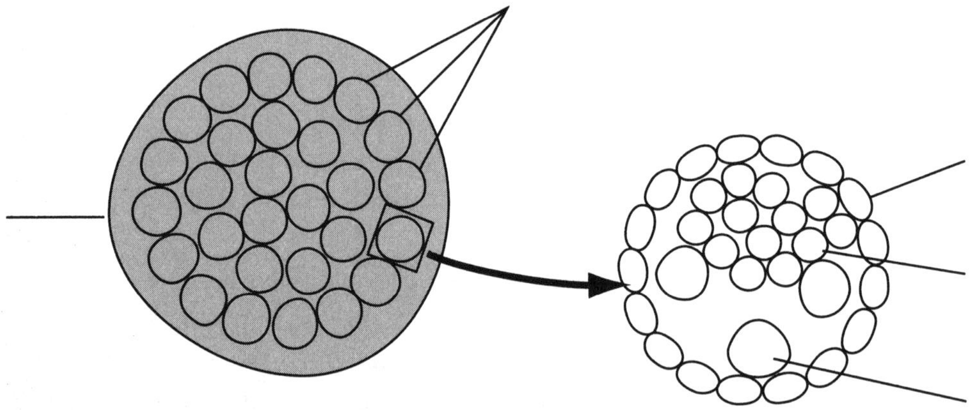



2. Label the parts of the dicot stem on the following diagram and color the parts as outlined in Procedure 6.

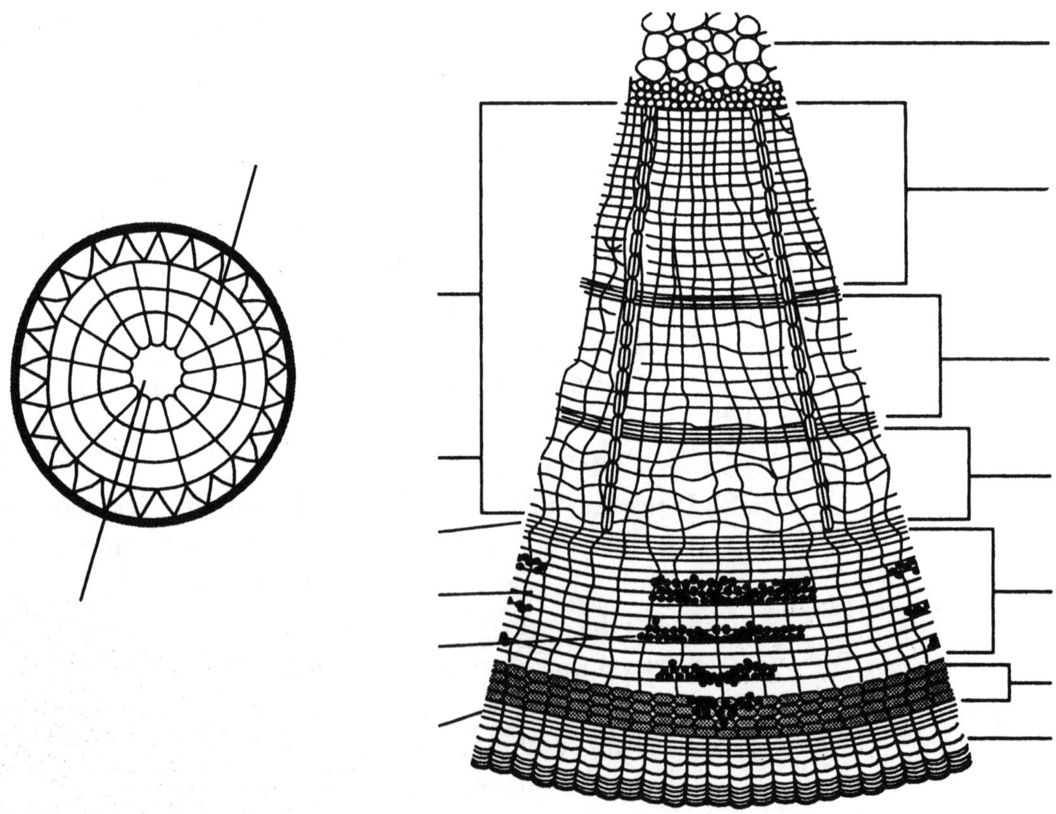

## CONCLUSIONS

**1.** What composes most of the cells in a monocot stem and in a dicot stem?

______________________________________________

______________________________________________

**2.** What are the functions of the following in a plant stem?

a. xylem ______________________________________________

______________________________________________

g. phloem ______________________________________________

______________________________________________

h. cambium ______________________________________________

______________________________________________

i. epidermis ______________________________________________

______________________________________________

**3.** How can you determine the age of a dicot? How old is the one you observed under the microscope?

______________________________________________

______________________________________________

______________________________________________

______________________________________________

**4.** What functions do stems serve?

______________________________________________

______________________________________________

______________________________________________

______________________________________________

**5.** If you girdle a tree (i.e., cut a ring around the tree through the bark), why does the tree die?

______________________________________________

______________________________________________

______________________________________________

______________________________________________

## OPTIONAL INTERNET SUPPLEMENT

Colorado State University Ag Extension (http://www.ext.colostate.edu): Search—plant structures/stems.

This Web site provides information about plant stem structures and functions.

University of Rhode Island (http://www.uri.edu/cels/bio/plant_anatomy/images.html):

Search—Anatomy of Stems. This Web site provides images of microscopic plant stems.

© Fedorov Oleksiy/Shutterstock.com

# EXERCISE 19

## Turn Over a New Leaf—Leaf Anatomy

### PURPOSE

In this lab, students become familiar with the structures and functions of a typical plant leaf.

### MATERIALS NEEDED

- Leaf section slide
- Compound microscope
- Set of colored pencils

### PROCEDURE

**1.** Study the functions of the leaf parts described here and identify them on the labeled diagram of a leaf cross-section.

a. **cuticle**—a very thin, non-cellular layer of waxy material found on both the top and bottom of the leaf.

b. **upper epidermis**—a single layer of heavy, walled cuboidal cells that serve as protection for the top part of the leaf.

c. **lower epidermis**—a single layer of heavy, walled cuboidal cells that serve as protection for the lower part of the leaf.

d. **chloroplasts**—the bodies in a leaf containing chlorophyll, that imparts the green color to leaves.

e. **mesophyll**—the leaf tissue between the upper and lower epidermis; contains palisade cells, air space, and a spongy layer.

f. **palisade cell**—long, rectangular cells composing a layer one- or two-cells in thickness just below the upper epidermis; chloroplasts are found in these cells.

g. **air spaces**—spaces of air found between the spongy layer cells.

h. **spongy layer**—layer of loosely packed cells several cell layers thick; chloroplasts are found in this layer.

i. **stoma (Stomata)**—openings found in the lower epidermis of a leaf that moisture and gases pass through.

j. **guard cells**—crescent-shaped cells that stand out from other epidermal cells because of their thickened walls and because they contain chloroplasts. Guard cells control the moisture content of the leaf and the exchange of gases.

k. **vascular bundle**—conducts water to the cells of the leaf and carries manufactured food from the leaf to the other parts of the plant. The vascular bundle is really a vein of the leaf.
l. **xylem**—thick-walled, upper-conducting cells in the vascular bundle.
m. **phloem**—thin-walled, lower-conducting cells in the vascular bundle.

**2.** Examine a leaf under a microscope.
  a. place a prepared slide of a leaf cross-section on the stage. Center it over the light opening and fasten it in place with the stage clips.
  b. focus with the low-power lens and observe the leaf parts. You may want to refer to the diagram in Observation 1.
  c. label the leaf parts on the diagram in Observation 1.
  d. find a vein of the leaf and switch to high power.
  e. observe the xylem and phloem.

**3.** Color the leaf parts in the diagram in Observation 1 as follows:
  a. cuticle—orange
  b. upper epidermis and lower epidermis—yellow
  c. palisade cells—pink
  d. air space—white
  e. spongy layer—dark green
  f. stoma—light green
  g. guard cells—red

## OBSERVATIONS

**1.** Review the definitions in one of the procedures above and refer to Figure 15-18 in your textbook. Label the parts of the leaf in the following diagram, and color the parts outlined in procedure 3.

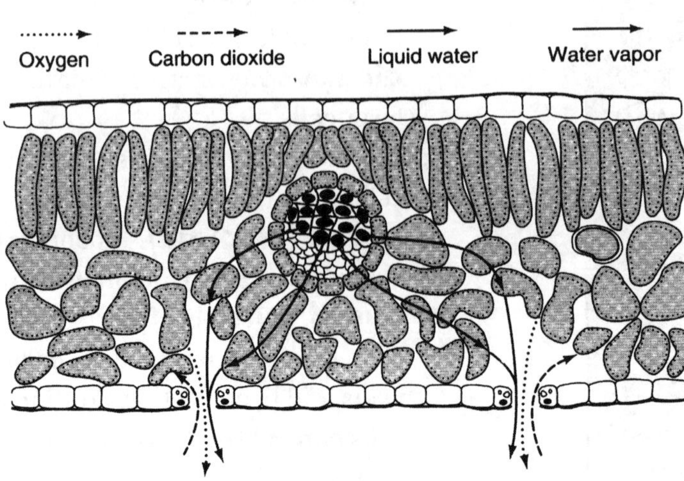

## CONCLUSIONS

**1.** What is the main function of the leaf in a plant?

**2.** What function do chloroplasts serve?

**3.** Can you name some plants whose leaves would not contain chlorophyll?

**4.** What is the function of each of the following leaf parts?

a. Epidermis

b. stomata

c. guardcells

d. cuticle

e. chloroplasts

## OPTIONAL INTERNET SUPPLEMENT

Colorado State University Ag Extension (www.cmg.colostate.edu): Search—plant structures/ leaves. This Web site provides information about leaf structures and functions.

© Anne Kitzman/Shutterstock.com

# EXERCISE 20

## He Loves Me, He Loves Me Not—Flower Anatomy

### PURPOSE

This exercise enables you to become familiar with the structures of a plant's flower and understand how they function.

### MATERIALS NEEDED

- Taraxacum young bud slide
- Taraxacum older bud slide
- Mature anther slide
- Pollen germination slide
- Compound microscope
- Set of colored pencils

## PROCEDURE

**1.** Study the following parts and functions while observing the labeled diagram. Refer to Figure 15-19, parts of a flower, in your textbook.

a. **stamen**—collectively, the male reproductive parts of a plant flower: the filament, anther, and pollen.
b. **filament**—long stalk that supports the anther. The anther is attached to the end of the filament.
c. **anther**—the male sex organ of a plant flower; manufactures the pollen.
d. **pollen**—the male reproductive cell of a plant flowers.
e. **Pistil**—collectively, the female reproductive parts of a plant flower: the stigma, style, ovary, and ovules.
f. **stigma**—the center part of the plant flower; receives the pollen from the anther.
g. **style**—the pollen enters the stigma and travels down the style to the ovary.
h. **ovary**—the female sex organ of the plant's flower. Ovules are produced here.
i. **ovules**—the female reproductive cells of a plant; they are fertilized by the pollen and eventually ripen into seeds.
j. **perfect flower**—has all of the parts mentioned in 1a through 1i.
k. **imperfect flower**—at least one but possibly more of the parts mentioned in 1a through 1i are missing.
l. **petals**—parts of a flower producing a scent that acts as an attractant to insects and other pollinators.
m. **sepals**—serve as protection for the developing flower.

2. Place the young bud slide under the microscope and view it carefully. Draw the young bud in Observation 1.
3. Place the older bud slide under the microscope and view it carefully. Compare it to the young bud slide. Record any differences in Observation 2.
4. Place the mature anther slide under the microscope and view it carefully. Record what you see in Observation 3.
5. View the pollen germination slide. Make a drawing of your observation in Observation 4.
6. Complete the diagram in Observation 5.

## OBSERVATIONS

1. Draw the young bud as seen under the microscope.

2. What differences were observed between the young bud and the older bud?

___

___

___

3. Draw the mature anther as seen under the microscope.

**4.** Draw pollen germination as viewed under the microscope.

**5.** Use your textbook to label the parts of the following diagram.

**Parts of a Flower**

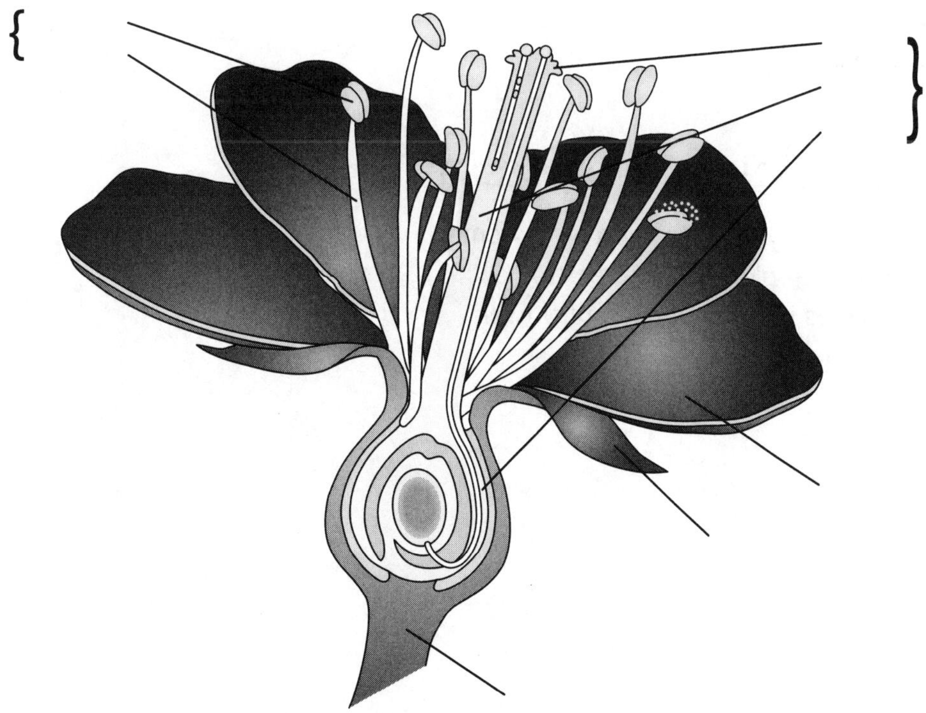

## CONCLUSIONS

**1.** Identify the functions of the following flower parts.

a. filament ____________________
b. anther ____________________
c. stigma ____________________
d. ovary ____________________
e. petals ____________________
f. sepals ____________________

**2.** Explain the difference between a perfect and an imperfect flower. ____________________

____________________

____________________

____________________

**3.** Color the parts of the flower diagram in Observation 5 as follows:

a. anther—green
b. filament—light blue
c. pollen —pink
d. stigma—purple
e. style—brown
f. ovary—orange
g. ovules—dark blue
h. petals—yellow
i. sepals—red

## OPTIONAL INTERNET SUPPLEMENT

Colorado State University Ag Extension (www.cmg.colostate.edu): Search—plant structures/flower. This Web site provides information about flower structures and functions.

© Roger Swift/Shutterstock.com

# EXERCISE 21

## Under the Knife—Flower Dissection

### PURPOSE

The development of fruit and seed begins with the flower for angiosperms. A complete flower differs from a perfect flower in that the complete flower contains all the floral whorls, whereas a perfect flower needs only to have a stamen and pistil.

### MATERIALS NEEDED

- A complete flower
- Dissecting probe (teasing needles)
- Scissors
- Scalpel (optional)
- Tweezers/forceps (optional)
- Dissecting scope or zoom lens 18 to 108 mm

## PROCEDURE

1. Pull the corolla from the flower, petal-by-petal, and record in Table 21-1 the number of petals.
2. Pull the calyx from the flower, sepal-by-sepal, and record in Table 21-1 the number of sepals.
3. Observe the stamens and pistil. Record the location of the pistil in the Table 21-1, then record the location of the stamens.
4. Dissect the remaining parts of the flower lengthwise down the middle.
5. Pull out one stamen and sketch the parts of the stamen (2 parts). Make your sketch in the labeled boxes for "Anther" and "Filament" in the Observations section.
6. Place a drop of water onto a microscope slide.
7. Holding a stamen with the tweezers or forceps over the drop of water on the microscope slide, gently tap the head of the stamen (the anther) with the dissecting probe. Pollen will fall from the anther to the drop of water. Examine the pollen using either a dissecting scope or zoom lens.

## OBSERVATIONS

1. Sketch a few pollen grains in the box labeled "Pollen grain."
2. Pull out the pistil and draw the parts of the pistil (3 parts). Make your sketches in the boxes labeled "Stigma," "Style," and "Ovary."
3. In the box labeled, "Ovules," sketch the cup-shaped ovary that holds the ovules.

| Dissection of a Complete Flower | |
|---|---|
| Number of petals | |
| Number of sepals | |
| Location of pistil | |
| Location of stamen | |



| **Anther** | **Filament** | **Pollen grain** |
|---|---|---|
| | | |

| **Stigma** | **Style** | **Ovary** |
|---|---|---|
| | | |

| **Ovules** | **Pistil** |
|---|---|
| | |

## CONCLUSIONS

1. What would make this flower an incomplete flower?

2. What would make this flower an imperfect flower?

## OPTIONAL INTERNET SUPPLEMENTS

Marietta College Department of Biology and Environmental Science (www.marietta.edu): Search "from flowers to fruits."

This Web site provides good information that will increase your understanding of the process that turns flowers into fruit.

© italianestro/Shutterstock.com

# EXERCISE 22

## Free Gas Made Here—Photosynthesis

### PURPOSE

One of the by-products of plant photosynthesis is the production of oxygen. In this exercise, you will be able to see how light intensity affects the amount of oxygen a plant will produce.

### MATERIALS NEEDED

- Calomba plant or other water plant
- Water
- Rectangular glass container
- Glass test tube
- Test tube clamp
- Grow light
- Napkin

### PROCEDURE

1. Set up the equipment for the experiment as shown above. Make sure your test tube is clamped to something to hold it in place. Fill a glass rectangular container with pure water. Place a grow lamp on the opposite side of the test tube containing the plant. Note: Keep the water plant in a dark place for a few hours before this experiment.
2. Place the grow lamp two feet from the test tube. Watch for a stream of small bubbles coming from the cut end of the water plant. These bubbles are oxygen that is being produced due to photosynthesis.
3. Count the number of bubbles the plant is producing for one minute. Repeat this process a total of three times. Find the average or mean for the bubble count. This is the rate of oxygen production for this experiment. Record your data in the table shown in Observation 1.
4. Change the distance to one foot from the test tube and repeat the steps in #2. Record your data in the table shown in Observation 1.

5. Place a thin napkin in between the glass container and the light. This will simulate cloud cover. Repeat step #2. Record your data in the table shown in Observation 1.
6. Plot your data on the graph in Observation 2. The values for the factor being varied should be on the x-axis, and the mean bubble rate in bubbles per minute should be on the y axis.

## OBSERVATIONS

1. Record your observations per Procedure steps 3-5.

| Variable | 1st | 2nd | 3rd | Mean |
|---|---|---|---|---|
| 2′ | | | | |
| 1′ | | | | |
| Cloud Cover | | | | |

2. Plot your data in the graph shown here per Procedure 6.

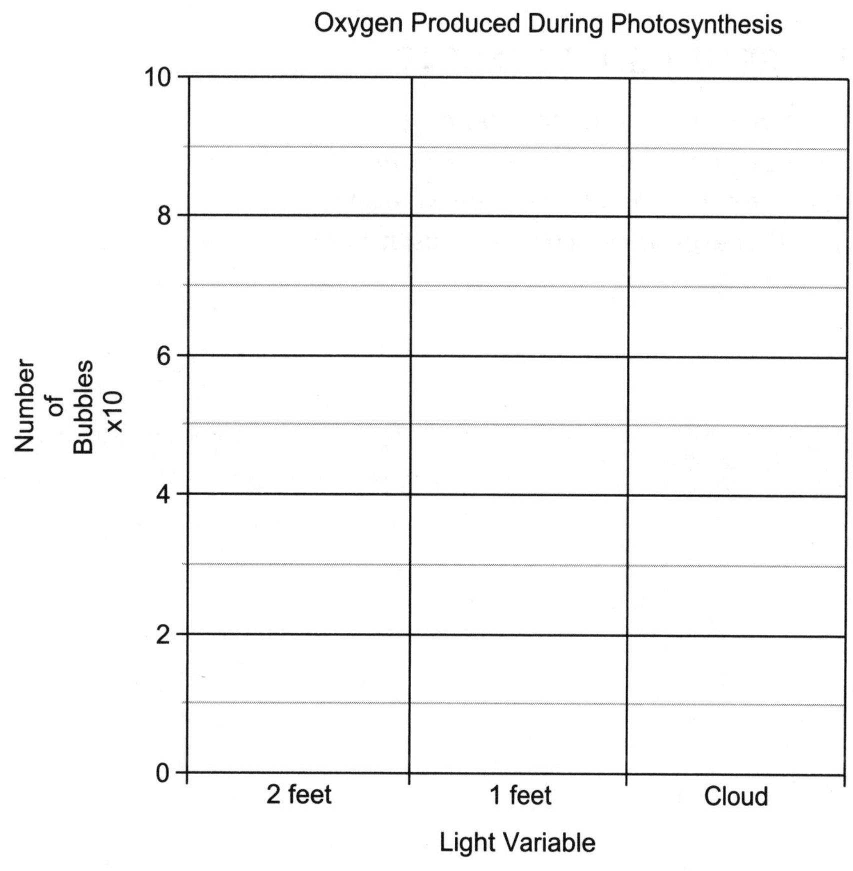

## CONCLUSIONS

1. Which variable caused the most oxygen to be released?

______________________________________________

______________________________________________

______________________________________________

2. How did the light intensity affect the rate of photosynthesis?

______________________________________________

______________________________________________

______________________________________________

3. What do you think was the purpose for the rectangular container of water? Could it have been round? Explain.

______________________________________________

______________________________________________

______________________________________________

## OPTIONAL INTERNET SUPPLEMENT

Nuffield Foundation (www.nuffieldfoundation.org):
Search practical-biology/photosynthesis/investigating factors affecting the rate of photosynthesis. This Web site provides ideas for further experiments with photosynthesis.

Source: Nuffield Foundation

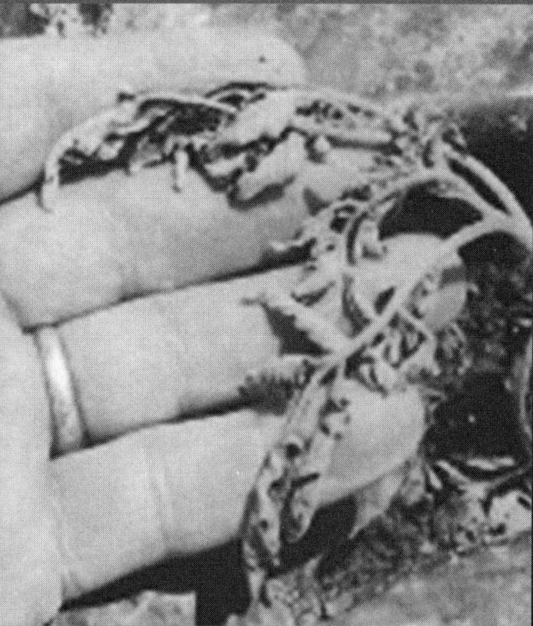
Courtesy of DeVere Burton

# EXERCISE 23

## Transpiration—Nature's Big Gulp

### PURPOSE

Transpiration is the process that water is pulled through a plant, starting at the roots into the body of the plant and then evaporating into the atmosphere. In this exercise, you will be able to observe this process.

### MATERIALS NEEDED

- 2 geranium leaves
- 2 small jars (baby food or Mason jars, or 2 beakers)
- Cardboard (10 cm square with a hole about the size of a hole-punch in center)
- Clay, Silly Putty, or petroleum jelly
- Water

### PROCEDURE

1. Roll the geranium leaf into a tube length wise. Place the top portion of the leaf in the hole of the cardboard square. Seal the bottom of the hole with clay or Silly Putty or petroleum jelly. Avoid getting any of the sealing medium on the top of the leaf.
2. Fill one small jar with water.
3. Place the cardboard with the leaf over the jar of water. Make sure most of the leaf is in the water, but not so much that the cardboard comes into contact with the water.
4. Cover the leaf with a second jar. Place the leaf in a well-lighted area.
5. Observe the inside of the top jar after 24 hours.

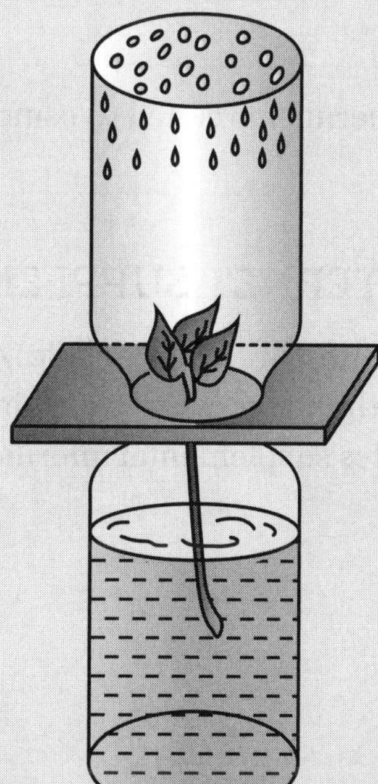

## OBSERVATION

1. Sketch your set-up after you have completed it in the box labeled "Start."
2. Sketch the set-up again after 24 hours in the box labeled "24 hours."
3. What changes took place?

______________________________________________

______________________________________________

______________________________________________

______________________________________________

| Start | 24 hours |
| --- | --- |
| | |

## CONCLUSIONS

1. Define equilibrium. ______________________________________________

______________________________________________

2. What is osmosis? ______________________________________________

______________________________________________

3. How is this experiment related to osmotic force or turgor? ______________________________________________

______________________________________________

## OPTIONAL INTERNET SUPPLEMENT

USGS Georgia Water Resource Center (http://ga.water.usgs.gov): Search—the water cycle/ transpiration. This Web site provides supplemental information about transpiration.

USDA/ADS #K1657-16. Photo by Jack Dykinga

# EXERCISE 24

## Reap What You Sow—Seed Anatomy

### PURPOSE

The purpose of this exercise allows you to become familiar with the structures and functions of a seed.

### MATERIALS NEEDED

- Seed of a plant slide
- Compound microscope
- Several fresh bean seeds
- Magnifying lens
- Scalpel

## PROCEDURE

**1.** Study the following parts of a seed:

a. **fruit coat**—transparent outer layer of the seed.

b. **endosperm**—makes up most of a seed and is composed of an opaque part, that contains starch, and a transparent part, composed of protein. These are the reserve foods the embryo uses when it begins to grow.

c. **seed coats**—layers that surround the endosperm and serve to protect the seed's internal parts.

d. **embryo**—Externally, the part of the seed marked by a slight depression; internally, the part of the seed that grows into a new plant. Composed of the plumule, the plumule sheath, the cotyledon, the radicle, and the radicle sheath.

e. **cotyledon**—makes up most of the embryo. The cotyledon's infolded edges almost enclose the plumule. Collectively, the cotyledons are the seed leaves.

f. **plumule**—the primary bud of the seed embryo.

g. **plumule sheath**—made up of several immature leaves, will become the first leaves of the plant.

h. **radicle**—becomes the root system of the plant.

i. **radicle sheath**—sheath of protection that surrounds the radicle.

j. **point of attachment**—where the seed is attached to the mature plant.

2. Place the slide of the corn seed under the microscope and identify the parts. Label these parts on the diagram in Observation 1.
3. Use the scalpel to cut one bean seed as shown in this diagram. Record what you find inside in Observation 2.

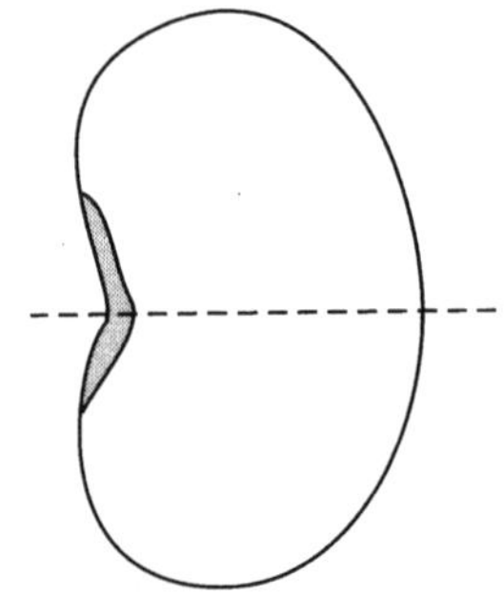



4. Cut another bean seed as shown here. Record what you see in Observation 3.

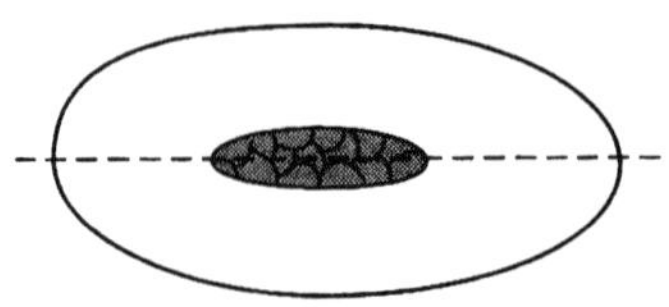



## OBSERVATIONS

1. Using the Web site at the end of this unit or your textbook materials, label the parts of the seed in this diagram.

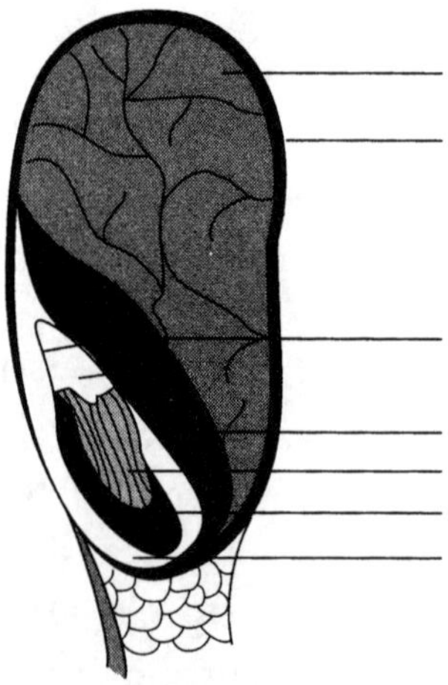



2. Draw what you observed inside the bean seed cut in Procedure 3.

**3.** Draw what you observed inside the bean seed cut in Procedure 4.

______________________________________________

______________________________________________

## CONCLUSIONS

**1.** What is the function of the largest part of a seed?

______________________________________________

______________________________________________

______________________________________________

______________________________________________

**2.** What are the functions of the following seed parts?

a. plumule ______________________________________________

______________________________________________

b. radicle ______________________________________________

______________________________________________

c. embryo ______________________________________________

______________________________________________

d. seed coat ______________________________________________

______________________________________________

**3.** What is the function of seeds in the life cycle of a plant?

______________________________________________

______________________________________________

______________________________________________

______________________________________________

## OPTIONAL INTERNET SUPPLEMENT

Colorado State University Ag Extension (http://www.ext.colostate.edu): Search—plant structures seeds. This Web site provides information about flower structures and functions.

© iStockphoto/Dimitris Stephanides

# EXERCISE 25

## Come Out; Come Out Wherever You Are—Germination

### PURPOSE

This lesson guides you to determine how differences in environmental conditions affect seed germination.

### MATERIALS NEEDED

- Water
- Aluminum foil
- Beaker
- Hot plate
- Freezer
- Seed-Gro Kit

## PROCEDURE

**1.** Moisture experiment
   a. place a piece of dry aluminum foil in a dry place.
   b. place five seeds on the foil or paper. *Do not water.*
   c. place one of the foam germination pads in the petri dish and water as directed by your instructor.
   d. place three seeds on a foam pad, and add a few drops of water directly on the seeds to hasten germination.
   e. place the cover on the germination chamber to reduce moisture loss and keep out unwanted materials.
   f. observe the seeds for a 7-day period. Record your observations on the chart in Observation 1.

**2.** Air (oxygen) experiment
   a. fill a beaker approximately half full with water.
   b. place five seeds in the beaker.
   c. place one of the foam germination pads in the petri dish and water as directed by your instructor.
   d. place three seeds on the foam pad, and add a few drops of water directly on the seeds to hasten germination.
   e. place the cover on the germination chamber to reduce moisture loss and keep out unwanted materials.
   f. observe the seeds for a 7-day period. Record your observations on the chart in Observation 2.

**3.** Temperature experiments
   a. prepare two packets, each containing five seeds wrapped in aluminum foil.
   b. place one packet in the freezer for one week.
   c. place the other packet in boiling water for 10 minutes.

d. prepare three germination chambers by placing the foam germination pads in the petri dishes and watering as directed by your instructor.
e. label the germination chambers "hot," "cold," and "normal."
f. plant the seeds boiled in water in the chamber labeled "hot." Place the freezer-conditioned seed in the chamber labeled "cold." Add the five untreated seeds in the chamber labeled "normal."
g. observe the seeds for a 7-day period.
h. record your observations on the chart in Observation 3.

**4.** Light experiment
a. prepare two germination chambers by placing the foam germination pads in the petri dishes and watering as directed by your instructor.
b. plant three seeds of the same kind in each chamber.
c. wrap one chamber in aluminum foil, making sure no light can get into the chamber.
d. observe the seeds every day for one week. Try to let no light (or as little as possible) get to the seeds in the aluminum-foil-wrapped chamber when you observe them.
e. record your findings on the chart in Observation 4.

## OBSERVATIONS

**1.** Record your moisture observations in the following chart:

| | Day 1 | Day 2 | Day 3 | Day 4 | Day 5 | Day 6 | Day 7 |
|---|---|---|---|---|---|---|---|
| Germination Chamber Seeds | | | | | | | |
| Foil-Wrapped Seeds | | | | | | | |

**2.** Record the results of the air experiment in the following chart:

| | Day 1 | Day 2 | Day 3 | Day 4 | Day 5 | Day 6 | Day 7 |
|---|---|---|---|---|---|---|---|
| Germination Chamber Seeds | | | | | | | |
| Seeds in Beaker of Water | | | | | | | |



**3.** Record the results of the temperature experiment in the following chart:

| | Day 1 | Day 2 | Day 3 | Day 4 | Day 5 | Day 6 | Day 7 |
|---|---|---|---|---|---|---|---|
| Normal Seeds | | | | | | | |
| Cold Seeds | | | | | | | |
| Hot Seeds | | | | | | | |



**4.** Record the results of the light experiment in the following chart:

| | Day 1 | Day 2 | Day 3 | Day 4 | Day 5 | Day 6 | Day 7 |
|---|---|---|---|---|---|---|---|
| Germination Chamber Seeds | | | | | | | |
| Foil-Wrapped Seeds | | | | | | | |

## CONCLUSIONS

1. What part does moisture play in seed germination?

2. What happened to the seeds that were placed in the beaker of water? What caused this?

3. What effect did heat have on the germination of the seeds?

4. What effect did cold temperature have on the germination of the seeds?

5. How did the absence of light affect the germination of the seeds?

6. What conditions are necessary for proper seed germination?

## OPTIONAL INTERNET SUPPLEMENT

Indiana University Department of Biology (http://plantsinmotion.bio.indiana.edu): Enter site and select- Germination/corn. This Web site shows time lapse seed germination.

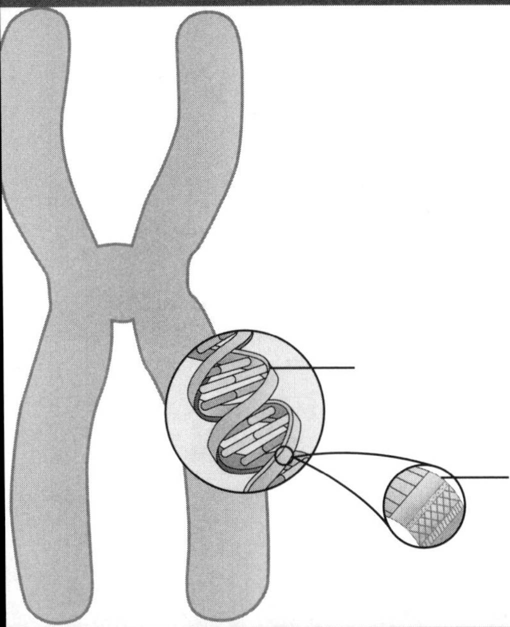



# EXERCISE 26

## Let's See What You're Made of—DNA Extraction

### PURPOSE

This lab will allow you to extract DNA and actually see what millions of strands of DNA look like.

### MATERIALS NEEDED

*(per student pair):*

- 1 plastic quart-size bag
- 1 strawberry
- 20 mL DNA extraction buffer
- Filter paper
- Ice cold 70–90 percent ethanol or isopropyl alcohol
- 1 Test tube or champagne flute
- Funnel (optional)
- Inoculation loop or coffee stir straw
- Microscope
- Slides and Coversheets

### PROCEDURE

1. Place one thawed strawberry in the plastic bag and squeeze until it is a uniform puree.
2. Add 20 mL of extraction solution. Extraction solution can be made as follows; 900 mL water, 100 mL shampoo and 2 tsp salt. Mix gently. This is enough solution for the class. Zip the bag closed.
3. "Moosh" the mixture of strawberry and buffer solution (without creating bubbles).
4. Fold filter paper or paper towel into a half circle, then a quarter circle, and open it to form a cone.
5. Fill the test tube (about 2 inches) or champagne flute to just above the stem with ice-cold alcohol. Place the filter paper cone into the test tube or champagne flute so half of the cone is on the inside of the tube and half on the outside of the tube.
6. Fill the paper towel cone with the strawberry solution.
7. As the strawberry mixture filters through the cone and comes in contact with the alcohol, the DNA will form ribbons and then coagulate at the top of the alcohol.
8. Use the straw to spool the DNA.
9. Prepare a wet mount slide and observe the DNA under a microscope.

## OBSERVATIONS

1. Did the DNA strands go to the bottom of the alcohol and then grow upward, or did the strands begin forming as soon as the strawberry solution made contact with the alcohol?

______________________________________________

______________________________________________

2. What color are the strands of DNA? ______________________

3. Did your group extract more or less DNA strands than other groups in the class?

______________________________________________

______________________________________________

4. Why do you think your group extracted more or less DNA strands?

______________________________________________

______________________________________________

5. In the boxes below, sketch your observations of the DNA without magnification and with magnification.

| **Without Magnification** | **With Magnification** |
| --- | --- |
| | |

## CONCLUSIONS

1. If you tried to see a single thread from across the room, you probably would not be able to see it. But you would be able to see a thick rope made up of many thousands of threads twisted together. How is this similar to the DNA isolation you just performed?

______________________________________________

______________________________________________

______________________________________________

______________________________________________

**2.** How could you design this experiment to show that only living (or once living) organisms have DNA?

______________________________________________________________________________

______________________________________________________________________________

______________________________________________________________________________

______________________________________________________________________________

## OPTIONAL INTERNET SUPPLEMENT

University of Utah- Genetic Science Learning Center (http://learn.genetics.utah.edu): Search—virtual labs/DNA extraction/try it yourself! How to extract DNA from any living thing. This Web site explains the process of DNA extraction.

# SECTION 6

## Crop Science

Courtesy of DeVere Burton

# EXERCISE 27

## We Are Here, We Are Here, We Are Here—Soil Organisms

### PURPOSE

In this exercise you will learn to recognize the different organisms found in the soil that aid in decomposition of organic matter and the formation of humus.

### MATERIALS NEEDED

- What's in the Soil II—Microorganism Kit
- Compound microscope—300× or better
- Four 400-mL beakers
- Test tube rack
- Two test tubes
- Bunsen burner
- Handful of green leaves
- Soil samples
- Water
- Personal Safety Set

**SAFETY NOTE**

Always wear eye protection glasses and rubber gloves when handling specimens. Wash hands before leaving the laboratory.

### PROCEDURE

1. Study the information on soil organisms provided in the microorganism kit.
2. Collect a soil sample to be used in this experiment. Record the information regarding the collection site in Observation 1.
3. Label one test tube "broth." Add nutrient broth to the test tube, filling the tube approximately two-thirds full. Keep the tube aside.
4. Label one test tube "agar." Pour the agar into the test tube and set the tube at a slant to allow the agar to harden. (Hardening at a slant provides a greater surface for organism growth.)
5. Place soil from your soil sample into a cup and add a small amount of water. Swirl the cup gently to mix the soil and water. Add more water so that the cup is two-thirds full.
6. Use a clean eyedropper and add a drop of the soil–water mixture to the tube containing the nutrient broth. Seal the tube and swirl it gently to mix the contents. Place the test tube in the rack. Check the tube in 24 hours. Record your results in Observation 2.
7. Sprinkle a small amount of soil on the agar in the slanted test tube. Seal the tube and place the test tube in the rack. Check the test tube in 48 hours. Record the results in Observation 3.
8. Wash a small piece of green leaf in water. Rinse the leaf in alcohol to remove any organisms from its surface and rinse again in water.

   **WARNING:** Alcohol is flammable. Do not expose to heat or flame.

9. Add the leaf to the cup containing the soil–water and place the cup in a location that receives sunlight. Be sure you have labeled the cup with your name. Check the cup in 48 hours. Record your findings in Observation 4.
10. Place a drop of water from the nutrient broth mixture on a glass slide, add a drop of methylene blue, and place a coverslip over the slide. Place the slide on the microscope and view first under low power and then under high power. Record your findings in Observation 5.
11. Place a piece of the growth from the agar solution on a slide, add a drop of methylene blue, and place a coverslip over the slide. Place the slide on the microscope and view first under low power and then under high power. Record your findings in Observation 6.
12. With an eyedropper, remove a small amount of water from near the surface of the leaf in the cup of soil. Place a drop on a slide and cover with a coverslip. View the slide under the microscope and record your findings in Observation 7.

## OBSERVATIONS

1. Record the following information regarding the soil sample site:
   a. location of sample ____________________
   b. light conditions ____________________
   c. temperature conditions ____________________
   d. humidity conditions ____________________
2. Record your findings regarding the nutrient broth.

   ____________________

   ____________________

   ____________________
3. Record your findings regarding the agar test tube.

   ____________________

   ____________________

   ____________________

4. Record your observations regarding the soil water.

5. Draw or describe what you see on the slide of the nutrient broth and methylene blue.

6. Draw or describe what you see on the slide of the agar and methylene blue.

7. Draw or describe what you see on the slide of the soil water.

## CONCLUSIONS

1. What soil organisms were observed in this experiment?

2. Why are these organisms found near the soil surface?

3. What functions do these organisms serve in the soil?

4. Were any of the organisms you found plants? How can you tell?

**5.** What other type organisms are in the soil that can not be seen?

______________________________________________

______________________________________________

## OPTIONAL INTERNET SUPPLEMENT

Discovery Education, The Dirt on Soil (http://school.discoveryeducation.com/schooladventures/soil/): This Web site allows the user to go on a virtual tour and examine soils.

Courtesy of DeVere Burton

# EXERCISE 28

## A Good Start—Growing and Transplanting Vegetable Seedlings

### PURPOSE

The purpose of this lesson is to practice the proper methods of sprouting and transplanting seedlings.

### MATERIALS NEEDED

- Tomato seeds
- Seed-starting media
- Flats
- Spray bottle with water
- Trowel or spoon
- Plastic wrap
- Large containers (10 ounces each or larger)
- Plant-growing media
- Liquid fertilizer
- Ruler
- Toothpicks

### PROCEDURE

**1.** Study the following terms.

a. **flat**—a wooden or a plastic box with slotted bottom used to start seedlings

b. **germinate**—to sprout or begin to grow

c. **media**—materials used to start and grow seeds and plants

d. **seedlings**—young plants that have been germinated several days

e. **transplant**—to move plants from one growing location to another thus giving them more space in which to develop

**2.** Growing the seedlings

a. fill the sections of the flat with the seed-starting media almost to the top.

b. place something under the flat to catch any excess water.

c. use the spray bottle to mist the mixture until it is thoroughly damp.

d. place two tomato seeds in each compartment of the flat or in your assigned compartments if sharing a flat with a classmate.

e. use the trowel or spoon and sprinkle additional seed-starting media on top of the seeds no more than $\frac{1}{8}$ inch deep.

f. mist the mixture again.

g. wait 10 to 15 minutes and mist once more with the spray bottle.

h. place four toothpicks in a square on the outer edges of the flat. Place another toothpick in the center. Cover the flat with plastic wrap being careful not to cover it too tightly. The toothpicks will work as tent poles to hold the wrap off of the media.
i. place the flat in a nice warm spot. Check daily for moisture. Mist if necessary.
j. once the seeds sprout, they need light, either artificial or sunlight, for growth.
k. once the tomatoes have grown their first set of true leaves, they are ready to be transplanted to larger containers.

**3.** Transplanting the seedlings
a. place about 8 ounces of growing media in each container. (Containers should have drainage holes.)
b. use the trowel or spoon to gently pry up the tomato seedlings.
c. gently hold the seedling by its leaves (not stems) and carefully place it in its new container.
d. firm the soil around it with your fingers.
e. use the spray bottle to thoroughly moisten the soil in the container.
f. make sure the containers are set where they receive adequate light and keep the soil moist (not wet).
g. after a week, provide the plants with a diluted (one-fourth to one-half strength) mixture of liquid fertilizer.
h. plants may be transplanted to their final growing location in about 1 month.

## OBSERVATIONS

**1.** Record the progress of your seed growth in the chart that follows. Record height and number.

| Tomato Seeds' Progress | Day 2 | Day 4 | Day 6 | Day 8 | Day 10 | Day 12 |
|---|---|---|---|---|---|---|
| Number Sprouting | | | | | | |
| Height of Tallest | | | | | | |
| Height of Shortest | | | | | | |

2. Record the progress of your transplanted tomato plants for three weeks.

| Tomato Plants' Progress | Day 2 | Day 4 | Day 6 | Day 7 | Day 14 | Day 21 |
|---|---|---|---|---|---|---|
| Height of Tallest | | | | | | |
| Height of Shortest | | | | | | |



## CONCLUSIONS

1. What environmental conditions were necessary for the sprouting of the seedlings?

2. What environmental conditions were necessary for the growth of the seedlings after sprouting?

3. What was your germination rate? ***Hint:*** Number of seedlings ÷ Number of seeds planted × 100.

4. If the germination rate was not 100 percent, give a possible reason.

5. Did all plants that were transplanted survive? Give a possible reason.

## OPTIONAL INTERNET SUPPLEMENT

University of Illinois Extension, Watch Your Garden Grow (http://urbanext.illinois.edu/veggies/planting.cfm): This Web site shows the process oftransplanting vegetables.

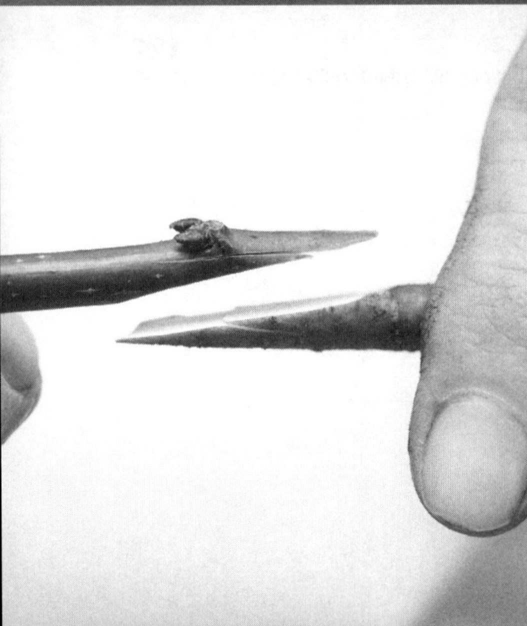
© Bork/Shutterstock.com

# EXERCISE 29

## Try That On for Size—Grafting Fruit Trees

### PURPOSE

In this exercise you will become familiar with the whip and tongue method of grafting fruit trees.

### MATERIALS NEEDED

- Knife
- Grafting wax
- Grafting tape
- Apple or other fruit stock—¼-to ½-inch diameter
- Scions—¼-to ½-inch diameter
- Leather gloves

### PROCEDURE

**1.** Study the following terms related to grafting.

a. **grafting**—uniting two different plants so they grow as one

b. **cambium**—thin, green, actively growing tissue located between the bark and wood of a plant

c. **grafting wax**—a pliable, sticky, waterproof material made of beeswax, resin, and tallow

d. **rootstock**—the lower portion of a graft that becomes the stem and roots of the new plant

e. **scions**—a short piece of shoot containing several buds that becomes the new top of a grafted plant.

**2.** Making the cuts

a. be sure the rootstock and scion are as close as possible to the same size.

b. using the knife, make the first cut on the scion below a bud. The length of the cut should be such that a smooth slanted angle is cut from 1½ to 2 inches long.

c. the rootstock should be cut next with the same length and slope as that of the scion so the two parts will fit together evenly.

d. on the scion, make a second cut beginning about one-third the distance from the tip. The cut should begin vertically but gradually become nearly parallel to the first cut surface. This is the tongue.

e. follow the same procedure as in the previous step with the rootstock. When finished, the scion and rootstock should fit together with the tongues interlocking.

3. Completing the graft
    a. make sure the cambium layers of the rootstock and scion are matched together on at least one side.
    b. push the scion and rootstock together tightly matching the surfaces as closely as possible. Be sure there is no overlap of scion and rootstock.
    c. wrap the graft with the grafting tape to keep it tight to prevent drying.
    d. apply grafting wax over the entire area that is wrapped with tape as uniformly as possible. This is an additional precaution to prevent drying.
    e. in both the taping and the waxing steps, be careful not to dislodge the aligned cambium layers of the scion and rootstock.

## OBSERVATIONS

1. In grafting the scion and rootstock together, why was it important to match the cambium layers together?

_______________

_______________

_______________

2. What could be the possible consequences of not wrapping and waxing the graft?

_______________

_______________

_______________

## CONCLUSIONS

1. Define the following terms:

    a. cambium _______________

    _______________

    b. grafting _______________

    _______________

    c. rootstock _______________

    _______________

    d. scion _______________

    _______________

**2.** Why is grafting used in the fruit and nut industries?

**3.** Is grafting an economical method of producing fruits and nut trees? Explain your answer.

## OPTIONAL INTERNET SUPPLEMENT

North Dakota State University (http://www.ag.ndsu.edu/ndsuag/): Search—grafting techniques. This Web site describes grafting techniques.

© Arkot/Shutterstock.com

# EXERCISE 30

## From Fries to Fuel—Making Biodiesel

### PURPOSE

With fuel costs at an all time high and exploration of new sources caught up in red tape, producers of biodiesel are finding their product in high demand. In this exercise, you will observe the process by which vegetable oil is used to produce fuel that can run your car, truck, or tractor. Due to chemical exposure, this exercise will be performed by your instructor.

### MATERIALS NEEDED

The complete materials list is provided in the instructor's manual.

- 1 liter *new* vegetable oil
- 4 grams substance A (sodium hydroxide or caustic lye)
- 250 mL substance B (is methanol or wood alcohol)
- 2 glass containers with lids (glass juice bottle)
- 1 funnel
- 4 qt pot
- Warming plate

### PROCEDURE

***Note:*** Do not attempt to make biodiesel at home! Wear safety gloves and eye protection glasses during this exercise, and make sure that all items of equipment are clean and dry.

1. On the warming plate, heat the oil to 55° C in the 4 qt pot and remove it from the heat.
2. Pour the 250 mL of substance B into the glass container and quickly replace the lid. Be careful not to breathe in any fumes.
3. Measure out four grams of substance A. Be sure not to let this substance come into contact with any moisture. Using a clean and dry funnel, add substance A to substance B in the glass container. Replace the lid.
4. Keeping container away from your face, swirl the mixture gently for 1 minute. Check to see if the granules are dissolved. If they are not, swirl again for 1 minute. Repeat this process until all of substance A has completely dissolved (this can take up to six times). The mixture will become very warm as the chemicals combine.
5. Check the blender to make sure all of the gaskets are in good condition and all parts are tightly fitted. Pour oil into the blender.
6. With the blender turned **OFF**, slowly add the A/B solution. Secure the blender lid and turn the blender on to a low speed. Allow the solution to mix for 20 minutes.
7. Transfer the solution to a clean glass jar. Within 20 minutes, you will begin to see a separation. After 24 to 36 hours, their will be two distinct layers. A clear yellow layer (biodiesel) and a thick brown layer (glycerin by-product).

- Thermometer
- Blender (Do not use with food again)
- Turkey baster (Do not use with food again)
- Protective eye wear
- Thick rubber gloves

## OBSERVATIONS

1. Describe the appearance of the three substances prior to mixing.

2. In your opinion, why do you think mixing substances A and B together produced heat?

3. Describe the appearance of the mixture soon after if was taken out of the blender.

4. Draw a picture of the mixture 20 minutes after it was taken from the blender. Label your picture.

## CONCLUSIONS

1. Did the mixture produce two distinct layers after it had time to settle? If not, what do you think happened?

2. List three benefits of producing biodiesel.

3. As a farmer, what crop/s could you produce for use in the biodiesel industry?

## OPTIONAL INTERNET SUPPLEMENT

Biodiesel.org: This Web site provides additional information about biodiesel.

Pennsylvania State University, College of Agricultural Sciences, Publications (http://extension.psu.edu/Publications): Search for the publication on "Biodiesel Safety and Best Management Practices for Small-Scale Noncommercial Use and Production" This publication provides safety information that should be used when manufacturing biodiesel.

© istockphoto/Dimitris Stephanides

# EXERCISE 31

## Deep Thoughts—Exploring Seed Planting Depths

### PURPOSE

This experiment allows you to observe the effects of planting depth on seed germination and crop production.

### MATERIALS NEEDED

- Six glass jars—pint size
- Soil
- Water
- Corn seeds

### PROCEDURE

**1.** Prepare the jars.

a. label the jars "1" through "6." Prepare the jars as follows.

1. jar 1—place 1 inch of loosely packed soil in the bottom of the jar and place one corn seed on each of the four sides of the jar next to the side where they can be seen; put 3 inches of loosely packed soil on top of the corn seeds.
2. jar 2—place 2 inches of loosely packed soil in the bottom of the jar and place one corn seed on each of the four sides of the jar next to the side where they can be seen; put 2 inches of loosely packed soil on top of the corn seeds.
3. jar 3—place 2½ inches of loosely packed soil in the bottom of the jar and place one corn seed on each of the four sides of the jar next to the side where they can be seen; put 1½ inches of loosely packed soil on top of the corn seeds.
4. jar 4—place 3 inches of loosely packed soil in the bottom of the jar and place one corn seed on each of the four sides of the jar next to the side where they can be seen; put 1 inch of loosely packed soil on top of the corn seeds.
5. jar 5—place 3½ inches of loosely packed soil in the bottom of the jar and place one corn seed on each of the four sides of the jar next to the side where they can be seen; put ½ inch of loosely packed soil on top of the corn seeds.

6. jar 6—place 4 inches of loosely packed soil in the bottom of the jar and place one corn seed on each of the four sides of the jar next to the side where they can be seen; do not cover.

b. care for the seeds as they germinate.
   1. moisten the soil thoroughly but not excessively.
   2. place the six jars where they will receive adequate light and warmth.
   3. monitor the progress of the seeds for three weeks, watering when needed.
   4. record your findings on the chart in Observation 1.

## OBSERVATIONS

**1.** Record the progress of your seeds in the following chart.

| Jar Number | Week 1 | Week 2 | Week 3 |
|---|---|---|---|
| 1 | | | |
| 2 | | | |
| 3 | | | |
| 4 | | | |
| 5 | | | |
| 6 | | | |



## CONCLUSIONS

**1.** Which planting depth (or depths) was/were most appropriate for the corn seed? Why?

______________________________________________

______________________________________________

______________________________________________

______________________________________________

**2.** What happened to the seeds that were planted the deepest?

______________________________________________

______________________________________________

______________________________________________

______________________________________________

**3.** Why is it important to crop production for seeds to be planted at the proper depth?

**4.** If planted in a field, what would probably have happened to the seeds in jar 6?

**5.** What conditions are necessary for proper seed germination?

## OPTIONAL INTERNET SUPPLEMENT

MBS Seed (http://www.mbsseed.com), Planting Chart:
This chart shows seed spacing and planting depth for optimal germination and growth.

Courtesy of DeVere Burton

# EXERCISE 32

## Does This Smell Good?—Storing Silage

### PURPOSE

This lab lets you observe how silage is produced and how different factors affect silage quality.

### MATERIALS NEEDED

- Freshly cut chopped grass
- Six 1-quart jars with tight-fitting lids
- Labels for the jars
- Pen
- Nails
- Hammer

### PROCEDURE

**1.** Prepare the forage
   - a. spread the grass out to dry for three days. If the grass has too much moisture, it will rot instead of fermenting.
   - b. label the jars "1" through "6."
   - c. prepare the grass as follows:
      1. jar 1—pack the grass tightly, fill with water, and cover the jar tightly.
      2. jar 2—pack the grass tightly, fill with water, cover the jar, and punch several holes in the lid.
      3. jar 3—pack the grass tightly, cover the jar tightly.
      4. jar 4—pack the grass tightly, cover the jar, and punch several holes in the lid.
      5. jar 5—pack the grass loosely, cover the jar.
      6. jar 6—pack the grass loosely, do not cover the jar.

**2.** Storing the silage
   - a. put the jars in a dark place.
   - b. check them every day for 3 weeks to observe the development of the silage.
   - c. compare the silage in your jars to the quality factors listed in the following table.
   - d. record your results in Observation 1.

| Comparing Silage Color and Odor | | | |
|---|---|---|---|
| **Color** | | **Odor** | |
| Natural green color. Some clovers may turn a darker color than normal. | Good | Clean, pleasant, vinegary or pickled smell. | Good |
| Light green or yellowish color; also may be slightly brown. | Fair | Slightly fruity, yeasty, or scorched smell. Corn with high moisture may have acidic smell. | Fair |
| Gray or whitish color (molding); black or dark brown (overheating). | Poor | Musty smell (molding); strong burnt odor (overheating); rotten odor (poor bacterial action). | Poor |



## OBSERVATIONS

Complete the following chart regarding the quality characteristics of the silage in your jars.

| Jar Number | Color | Odor | Other Observations |
|---|---|---|---|
| 1 | | | |
| 2 | | | |
| 3 | | | |
| 4 | | | |
| 5 | | | |
| 6 | | | |



## CONCLUSIONS

**1.** Which jar produced the best silage based on color and odor?

______________________________

______________________________

______________________________

**2.** What happens when forage is ensiled too wet?

______________________________

______________________________

______________________________

**3.** Why do you think silage is packed tightly when stored?

______________________________________________________________________

______________________________________________________________________

______________________________________________________________________

## OPTIONAL INTERNET SUPPLEMENT

Iowa Beef Center (http://www.iowabeefcenter.org): Search—Making Quality Corn Silage. This Web site provides information that will assist farmers in producing the best quality silage and offers a feeding schedule to produce the strongest gains.

USDA/ARS # K3653-12

# EXERCISE 33

## Stop Buggin' Me—Insect Collection

### PURPOSE

Students will learn to identify the types of insects that can cause harm to a field crop.

### MATERIALS NEEDED

- Hand lens 10–30× power, or a dissecting scope
- Sweep net (can be purchased from any biological educational catalog)
- Stapler
- Paper bags (large grocery store size)
- Glass jars with screw-top lids
- Ethanol or isopropyl alcohol (rubbing alcohol)
- Use of a freezer (optional)
- Glass dish or Petri dish
- Internet access

### Sweep Method: Importance to Industry

The use of a sweep net is a common method of monitoring insects in a field. In crops such as alfalfa and cotton, insects are collected and counted to estimate the population levels. Field monitoring is important to determine whether treatment is necessary.

In this exercise, the use of a sweep net is demonstrated in alfalfa. Insects of concern are the alfalfa weevil larvae (*Hypera postica*) or Egyptian alfalfa weevil (*Hypera brunneipennis*), alfalfa caterpillars (*Colias eurytheme*), beet armyworm (*Spodoptera exigua*), and western yellow-striped armyworm (*Spodoptera praefica*). Other insects, including predators and parasitic wasps (that are the good guys in this case) can also be detected.

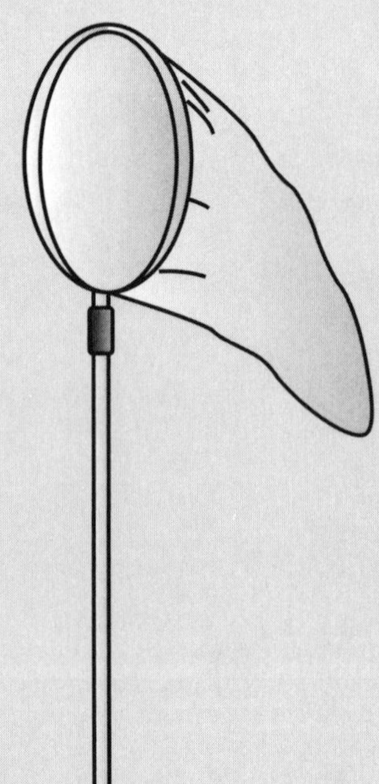



Insect collecting sweep net.

## PROCEDURE

1. A standard sweep net consists of a cone-shaped cloth bag approximately 2 feet deep fitted to a 15-inch wire loop and attached to a 26-inch handle. To sweep, tilt the rim of the net so the lower edge is 1 to 2 inches ahead of the upper edge of the rim. This allows the net to catch insects falling from the plant. The lower rim of the net should be kept about 10-inches below the top of the plant during sweeping. A single sweep is on a 180-degree arc taken during a step forward. At the end of this sweep, raise the net and reverse the direction of the swing. Swing the body from side to side during sweeping, making natural swings. Sweeps may be taken singly or consecutively.
2. To check a field, divide the field into four or more sections depending on the size of the field. Take five sweeps in each section and deposit the contents of the net into a paper bag and staple the bag closed. Place the paper bags from each section of the field into a freezer for 10 minutes. Open the bags and discard the plant material. Pour the insects into the glass jars and add enough alcohol to cover the insects. Seal the jars tightly to prevent the alcohol from evaporating. The alcohol preserves the samples for counting and identification at a later time.
3. To count and identify the insects, carefully pour off as much alcohol as possible without losing any insects. Place the insects in a shallow glass dish or Petri dish. If the insects float, there is too much alcohol remaining in the sample. Put the dish under the microscope and examine the sample. Identify the different insects and count the number of each species. To determine the average number per sweep, add the number of each species from each section of the field and divide by the total number of sweeps taken from the field.

## OBSERVATIONS

| | Average Number per Sweep | | | |
|---|---|---|---|---|
| **Insect:** | | | | |
| Number per Sweep | | | | |
| Number of Sweeps | | | | |
| Average Number per Sweep | | | | |

- The recommended treatment threshold for weevils is an average of 20 larvae per sweep.
- Caterpillars have a number of predators, parasites, and pathogens that may occur in a field that keep the populations below economically damaging levels. The treatment threshold for alfalfa caterpillars is an average count of 10 non-parasitized or disease-free caterpillars per sweep.
- Beet armyworm or western yellow-stripe armyworm need to be controlled when an average of 15 non-parasitized armyworms of more than ½ inch in length per sweep are found.

Control of these insects could be obtained with early harvest of the hay or with an insecticide. When economically practical, many growers choose to harvest early to minimize the killing of predators and parasites of aphids.

## CONCLUSIONS

1. Using the information above and the data you collected; do you believe a treatment plan should be established for this field? Why?

 ___________________________________________

 ___________________________________________

 ___________________________________________

2. Using the Internet, answer the following; what are the current practices used to control one of the insects you found?

 ___________________________________________

 ___________________________________________

 ___________________________________________

## OPTIONAL INTERNET SUPPLEMENT

Insect Identification.org: This Web site will help in insect identification.

# SECTION 7

# Ornamental Use of Plants

© Kleter/Shutterstock.com

# EXERCISE 34

## Keep 'em in the Dark—Plants and Light

### PURPOSE

This exercise allows you to determine how plants respond to different lighting conditions.

### MATERIALS NEEDED

- Six pots of pea seedlings
- Four light-tight boxes, each with a hole in one side
- One completely light-blocking box
- Light source

### PROCEDURE

1. Water the plants well so that you will not need to disturb them during this experiment.
2. Arrange the four light-tight boxes so their holes face different directions. For example, one box has its hole toward the east side, one has its hole toward the west side, one has its hole toward the north side, and one has its hole toward the south side.
3. Place a light source (a lamp will work) in the middle of the four boxes.
4. Place a pot of pea seedlings in each box.
5. Place one pot of pea seedlings where it gets light from all directions.
6. Place one pot of pea seedlings in the completely light-blocking box.
7. Leave the plants in the boxes for one week. At the end of the week, open the boxes and observe the plants.
8. Record the plants' responses to light on the chart in Observation 1.

## OBSERVATIONS

Complete the following chart regarding the plants' response to light.

| Lighting Condition | Plant Response | | |
|---|---|---|---|
| | Color of Plant | Direction of Growth | Length of Stem |
| Light from North | | | |
| Light from South | | | |
| Light from West | | | |
| Light from All Directions | | | |
| No Light | | | |



## CONCLUSIONS

1. What was the general effect of light versus darkness on the pea plants' growth? How can you explain this?

2. How did the color vary on the plants exposed to light versus the one kept in total darkness?

3. How did the plants move in response to the light?

## OPTIONAL INTERNET SUPPLEMENT

University of Michigan- Reaching Out (http://www.reachoutmichigan.org/): Search—Do Plants Need Sunlight?

This Web site explores a plants relationship with light.

© Anne Kityman/Shutterstock.com

# EXERCISE 35

## Chip Off the Old Block—Plant Cloning

### PURPOSE

Students will observe the methods of cloning plants to obtain uniformity in production.

### MATERIALS NEEDED

- Plant II Cloning: Propagation of African Violets Kit
- African violet plant
- Soap
- Bunsen burner
- Scalpel
- Water—Sterile
- Four to five large beakers
- African violet potting soil
- Ethyl alcohol
- Hot plate
- Laundry bleach
- Ten to twelve small plastic pots
- Personal Safety Set

**SAFETY NOTE**

Always wear eye protection glasses and rubber gloves when handling dissection specimens. Wash hands before leaving the laboratory.

### PROCEDURE

1. Study the sterile-technique procedures outlined in the cloning kit instruction guide. Follow these procedures when working with this equipment.
2. Follow the procedure outlined in the kit for preparation of shoot multiplication media.
3. Prepare the leaves for culture as described in the kit instruction guide.
4. Establish the shoot multiplication culture as outlined in the guide. Observe the plants every two weeks for three months. Record your findings on the chart in Observation 1.
5. Prepare the rooting medium according to the directions in the kit.
6. Transfer the plants to the rooting medium as outlined in the instruction guide found in the kit. Observe the plants for the next three weeks. Record your findings in Observation 2.
7. After numerous roots have appeared, carefully transfer the plants to pots containing African violet potting soil. Scrape all excess agar from the roots. Treat the young plants as you would any other plant.

## OBSERVATIONS

**1.** Record the plant progress from Procedure 4 in the following chart:

| 2 Weeks | 4 Weeks | 6 Weeks | 8 Weeks | 10 Weeks | 12 Weeks |
|---|---|---|---|---|---|
| | | | | | |



**2.** Record your findings from Procedure 6 in the following chart:

| 1st Week | 2nd Week | 3rd Week |
|---|---|---|
| | | |



## CONCLUSIONS

**1.** What type of plant reproduction is cloning?

______________________________________________

______________________________________________

______________________________________________

______________________________________________

**2.** What advantages can you think of for cloning plants? What disadvantages?

______________________________________________

______________________________________________

______________________________________________

______________________________________________

**3.** Do you think cloning is a productive method of reproduction in animals? Why?

______________________________________________

______________________________________________

______________________________________________

______________________________________________

**4.** What are some of the current controversies surrounding animal cloning?

## OPTIONAL INTERNET SUPPLEMENTS

British Broadcasting Company (http://www.bbc.co.uk/schools): Search—cloning plants.

This Web site describes the ways plants can be cloned.

# SECTION 8

# Animal Sciences

# EXERCISE 36

## You Eat Like a Pig—Simple Animal Digestion

### PURPOSE

Humans have a digestive system similar to that of a pig. In this exercise, you will examine and identify the parts and functions of a simple digestive system.

### MATERIALS NEEDED

- Textbook
- Rubber gloves
- Scalpel
- Dissecting microscope
- Personal Safety Set
- Preserved digestive tract of a pig or a fetal pig

**SAFETY NOTE**

Always wear eye protection glasses and rubber gloves when handling dissection specimens. Wash hands before leaving the laboratory.

### PROCEDURE

**1.** Study the following parts of the simple digestive system in Figure 26-9 in *Agriscience: Fundamentals & Applications, 6E* and locate them on the drawing.

**Alimentary Canal**

a. **mouth**—the beginning of the simple digestive tract; external opening through which food is ingested.
b. **pharynx**—passageway from the mouth to the esophagus. Muscles here help move the food into the esophagus.
c. **esophagus**—muscular passageway from the pharynx to the stomach. A flap of skin at the entrance to the esophagus called the epiglottis prevents food from entering the trachea.
d. **stomach**—gastric juices continue the breakdown of food in this part of the simple digestive system.
e. **small intestine**—food is further broken down for absorption here, as pancreatic juices and bile are added to the digesting food.
f. **caecum**—a blind gut; serves no function in swine but functions to help rabbits and horses digest forages.
g. **large intestine**—most moisture is squeezed from the undigested food here; also called the colon.
h. **rectum**—part of the alimentary canal where undigested food is held prior to release from the body.
i. **anus**—external opening of the digestive tract where undigested food is expelled from the body.

### Accessory Organs

a. **teeth**—serve to grind and pulverize food.
b. **salivary glands**—secrete digestive salivary juices (saliva), that begin the chemical breakdown of food.
c. **pancreas**—produces pancreatic digestive juices, that are released into the small intestine to help break down food.
d. **liver**—lobe-shaped organ that purifies the blood brought from the stomach, pancreas, and intestines. It also secretes bile, that is later added to the digestive juices in the small intestine.
e. **gallbladder**—stores some bile from the liver and empties bile into the large intestine as waste to be removed from the body.

**2.** Put on rubber gloves and safety goggles, then spread out the digestive tract so that all parts can be easily seen. Identify the following parts and label them on the diagram in Observation 1.
a. from the mouth, feed passes through the esophagus and into the stomach. Locate the esophagus.
b. follow the esophagus to the stomach. Trace the stomach toward the posterior of the tract until it joins the small intestine. The small intestine is the smaller tube that leads to the large intestine (the larger tube).
c. where the small intestine and large intestine join is the caecum, a blind gut.
d. follow the large intestine toward the posterior. There is a slight enlargement just before it reaches the anus. This is the rectum.
e. find the liver. It is a lobed, dark red-brown organ.
f. look for the gallbladder. It is a small, greenish sac embedded under the right lobe of the liver.
g. the pancreas is a small, pinkish, grainy organ found inside the bend made by the first section of the small intestine

**3.** Using a scalpel, remove a 2-inch section of the small intestine and cut it lengthwise. Wash its lining and examine the intestine section under a dissecting microscope. In Observation 2, draw what you see. The projections you see are called villi. They aid in the absorption of nutrients into the blood-stream from the small intestine.

## OBSERVATIONS

**1.** Label the parts of the simple digestive system on the following diagram.

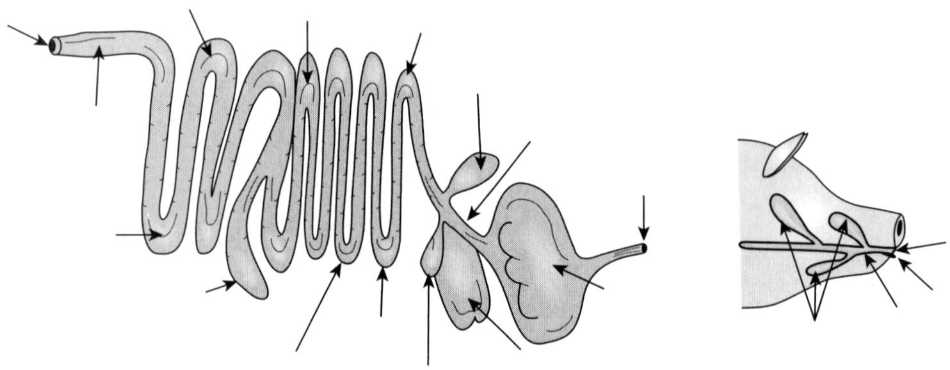



**2.** Draw the lining of a small intestine.

## CONCLUSIONS

**1.** What is the function of the digestive system?

______________________________________________

______________________________________________

______________________________________________

**2.** Match the functions (given below) with the structures of a monogastric digestive system.

a. Squeezes moisture from the undigested food ______________________

b. Produces bile ______________________

c. Grinds and pulverizes the food ______________________

d. Muscular passageway from the pharynx to the stomach ______________________

e. Accessory organ that is found in the first bend of the small intestine and produces digestive juices to aid in digestion ______________________

**3.** Which part of the digestive system digests roughages in horses and rabbits?

______________________________________________

______________________________________________

______________________________________________

**4.** What purpose do villi serve in the small intestine?

______________________________________________

______________________________________________

______________________________________________

## OPTIONAL INTERNET SUPPLEMENT

Canadian Museum of Nature (http://nature.ca/en/home): Search—Gallery Interactives: Build a Digestive System. Select-Human. This Web site is an interactive tutorial for a monogastric digestive system (pigs, sheep and humans etc.).

Courtesy of DeVere Burton

# EXERCISE 37

## It Takes Four—Ruminant Digestion

### PURPOSE

This exercise allows students to identify the parts and the functions of the ruminant digestive system.

### MATERIALS NEEDED

- Textbook
- Rubber gloves
- Scalpel
- Four jars
- Dissecting microscope
- Compound microscope
- Rumen bacteria slide
- Personal Safety Set
- Preserved digestive tract of a cow or sheep

**SAFETY NOTE**

Always wear eye protection glasses and rubber gloves when handling dissection specimens. Wash hands before leaving the laboratory.

### PROCEDURE

**1.** Study the following parts of the ruminant digestive system in Figure 26-8 of *Agriscience: Fundamentals & Applications, 6E* and locate them on the drawing.

**Alimentary Canal**

a. **mouth**—the beginning of the simple digestive tract; external opening through which food is ingested.

b. **pharynx**—passageway from the mouth to the esophagus. Muscles here help move the food into the esophagus.

c. **esophagus**—muscular passageway from the pharynx to the stomach. A flap of skin at the entrance to the esophagus called the epiglottis prevents food from entering the trachea.

d. **rumen**—the first compartment of a ruminant's stomach. It is sometimes called the paunch and makes up approximately 80 percent of the ruminant stomach. It acts as a large storage vat where food is agitated and fermented, and where digestion begins.

e. **reticulum**—this part of the ruminant stomach has a honeycomb-like lining that can catch and hold foreign objects that may be picked up by the animal. For this reason, it is sometimes referred to as the hardware stomach. It composes approximately 5 percent of the ruminant stomach. It, along with the rumen, has many bacteria and protozoa that aid in the breakdown of forages (roughages).

f. **omasum**—the third part of the ruminant stomach; grinds up food and may squeeze water from the food. It is sometimes called the manyplies and composes approximately 8 percent of the ruminant stomach.

g. **abomasum**—this part of the ruminant stomach is sometimes called the true stomach. It works similarly to the stomach in the simple digestive tract because the digestive juices are produced here. The abomasum composes approximately 7 percent of the ruminant stomach.
h. **small intestine**—food is further broken down for absorption here, as pancreatic juices and bile are added to the digesting food.
i. **caecum**—a blind gut; serves no function in ruminants but functions to help rabbits and horses digest forages.
j. **large intestine**—most moisture is squeezed from the undigested food here; also called the colon.
k. **rectum**—part of the alimentary canal where undigested food is held prior to release from the body.
l. **anus**—external opening of the digestive tract through which undigested food is expelled from the body.

**Accessory Organs**

a. **teeth**—serve to grind and pulverize food.
b. **salivary glands**—secrete digestive salivary juices (saliva), which begin the chemical breakdown of food.
c. **pancreas**—produces pancreatic digestive juices, which are released into the small intestine to help break down food.
d. **liver**—lobe-shaped organ that purifies the blood brought from the stomach, pancreas, and intestines. It also secretes bile, which is later added to the digestive juices in the small intestine.
e. **gallbladder**—stores some bile from the liver and empties bile into the large intestine as waste to be removed from the body.

**2.** Wearing rubber gloves and safety goggles, spread out the digestive tract so that all parts can be easily seen. Identify the following parts and label them on the diagram in Observation 1.

a. from the mouth, feed passes through the esophagus and into the rumen. Locate the esophagus.
b. follow the esophagus to the rumen. Trace the rumen to the reticulum. There is no real division between the rumen and reticulum.
c. next locate the omasum. Follow it until you find the abomasum.

d. follow the abomasum until it joins the small intestine. The small intestine is the smaller tube that leads to the large intestine (the larger tube).
e. where the small intestine and large intestine join is the caecum, a blind gut.
f. follow the large intestine toward the posterior. There is a slight enlargement just before it reaches the anus. This is the rectum.

3. Label the jars "rumen," "reticulum," "omasum," and "abomasum."
4. Cut the stomach open and take out some of the contents from each compartment. Put the contents in the jars labeled accordingly. Compare the digestion of food in each compartment. Record your findings in Observation 2.
5. Cut out a section of the inner lining of each part of the stomach. Wash each section thoroughly and observe it under the dissecting microscope. Draw what you see in Observation 3.
6. Place the rumen bacteria slide on the stage of the microscope and clip it down. Focus on low power and switch to high power. In Observation 4, draw the bacteria you see.

## OBSERVATIONS

1. Label the parts of the ruminant digestive system in the following diagram.

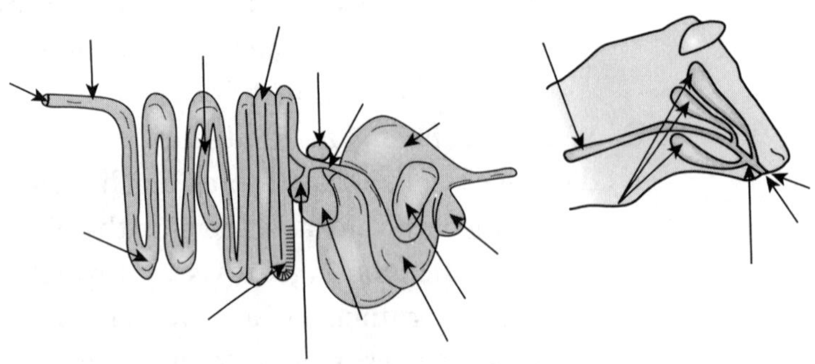

**2.** Describe the contents of each of the parts of a ruminant stomach.

a. rumen ______________________________

b. reticulum ______________________________

c. omasum ______________________________

d. abomasum ______________________________

**3.** Draw the inner lining of each part of the ruminant stomach as observed under the microscope.

| Rumen | Reticulum |
|---|---|
| | |
| Omasum | Abomasums |
| | |



**4.** Draw the bacteria as seen under the microscope.

## CONCLUSIONS

1. Identify the function of each of the parts of the ruminant stomach.
    a. rumen ______
    b. reticulum ______
    c. omasum ______
    d. abomasum ______
2. What type of feed can ruminants digest that simple stomachs cannot?
______
______
3. How do bacteria help digestion in the ruminant stomach?
______
______
4. Which part of the ruminant stomach performs the same basic function as does the simple stomach?
______
______
5. Define the term "ruminant."
______
______

## OPTIONAL INTERNET SUPPLEMENT

Canadian Museum of Nature (http://nature.ca/en/home): Search: Gallery Interactives: Build a Digestive System. Select: bison.

This Web site is an interactive tutorial for a ruminant digestive system.

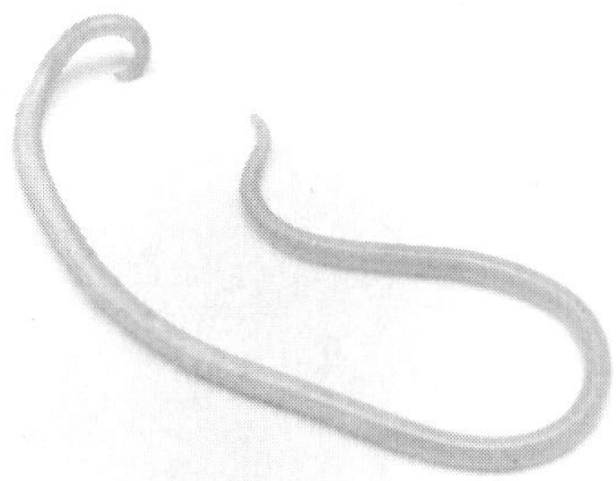

© Marcel Jancovic/Shutterstock.com

# EXERCISE 38

## Must Have Been Something I Ate—Internal Parasites

### PURPOSE

The purpose of this lab is to study the structure and life cycles of several economically important internal parasites.

### MATERIALS NEEDED

- Sheep liver fluke microscope slide
- Sheep tapeworm microscope slide
- Pig round worm microscope slide
- Compound microscope

### PROCEDURE

1. Study the information sheets for this exercise before continuing.
2. Examine the sheep liver fluke (*Fasciola hepatica* life history) slides under the microscope. Draw the liver fluke eggs in Observation 1.
3. Examine the sheep liver fluke infected snail tissue under the microscope.
4. Examine the sheep liver fluke infected sheep liver under the microscope. Compare the two slides of infected tissue in Observation 2.
5. Examine the sheep tapeworm slides under the microscope. Draw the tapeworm eggs in Observation 3.
6. Examine the pig roundworm slides under the microscope. Draw the pig roundworm eggs in Observation 4.
7. Study the life cycle charts of the three internal parasites.

### OBSERVATIONS

1. Draw the liver fluke eggs.

**2.** Compare the snail tissue damage to the liver damage. Draw or describe what you see.

**3.** Draw the tapeworm eggs.

**4.** Draw the pig roundworm eggs.

**5.** Complete the following life cycle charts by drawing arrows to indicate the cycle flow.

## CONCLUSIONS

**1.** What type of damage does each of the following cause?

a. liver fluke ______________________________

______________________________

b. tapeworm ______________________________

______________________________

c. roundworm ______________________________

______________________________

**2.** How do these three internal parasites differ in the damage they cause? How are they the same?

______________________________

______________________________

______________________________

**3.** What differences are there in the life cycles of these parasites? What similarities are there?

**4.** How are most internal parasites controlled?

## INFORMATION SHEET—LIVER FLUKE

The liver fluke is a parasite of cattle, sheep, goats, and humans. It is especially damaging to young animals.

The adult lives in the bile ducts where eggs are laid and passed down into the intestines and out in the feces. The eggs must land in water to hatch. The larvae that hatch from these eggs swim about seeking a snail. The snail is necessary for completing the life cycle. The larvae develop for a period in the snail, then emerge and encyst on plants along the water. Livestock eat the water plants and become infected. The young flukes pass to the intestines and burrow through the abdominal cavity and into the liver where they live principally on blood. Egg production begins approximately three months after entering the animal.

The fluke causes irritation, thickening of the bile duct, and fibrosis of the liver, making the animal unfit for human consumption.

The symptoms of liver flukes are anemia and weight loss. Highly infested animals may die.

Rotation of pastures and the use of water troughs help to control liver flukes. Chemical treatment kills the adult flukes in the animal. Control of snails will break the cycle but is hard to do.



## INFORMATION SHEET—TAPEWORM

There are several species of tapeworm that affect livestock. The broad tapeworm affects all classes of livestock and humans.

The adult lives in the small intestines, where it may reach a length of 10 or more feet. Segments of the tapeworm that contain eggs are continuously breaking off and passing out in the feces. The eggs are eaten by the orbatid mite that lives in the grass and weeds and serves as an intermediate host. The eggs develop in these mites, then are eaten by livestock, and hatch in the small intestines. They feed on animal foodstuff and grow to maturity. There is no physical damage to the host; rather tapeworms compete with the host for food.

Symptoms of tapeworms include unthriftiness (wasting away), loss of weight, diarrhea, and emaciation.

Tapeworms can be eliminated from their hosts via the use of chemicals.

## INFORMATION SHEET—ROUNDWORM

Roundworms are the most important group of internal parasites from an economic standpoint. There are many types of roundworm that affect almost every type of livestock. The roundworms of greatest concern are found in the digestive system, mostly in the stomach and intestines.

Roundworms in the adult stage live as blood-sucking parasites attached to the stomach wall. The eggs pass from the host to the feces and hatch into larvae in 15 to 20 days, depending on the temperature and humidity. The larvae crawl up blades of grass, are eaten by the animal, and travel to the stomach lining until they mature.

In the process of penetrating the stomach lining before maturing, they cause severe damage by reducing the digestion of nutrients by the host and producing poisons. Young, undernourished, or diseased animals are hardest hit.

The most common symptom is anemia, that can be indicated by a paleness of the gums and the whites of the eyes. In light infestations, the animal will have a dull hair coat, an unthrifty appearance, and will sometimes scour (have severe diarrhea). In severe infestations there will be persistent scouring, weight loss, anemia, prostration, and sometimes death.

Sanitation and pasture rotation are good control measures. Chemical de-wormers are used in treating infested animals.



## OPTIONAL INTERNET SUPPLEMENT

Food and Agriculture Organization of the United Nations (www.fao.org): Search initial surveys for determining the parasite species present.

This Web site can be used to identify internal parasites in livestock.

# EXERCISE 39

## We Are Doing What?—Fecal Examination for Parasites

### MATERIALS NEEDED

- Compound microscope
- Plastic bag (Ziploc® type)
- Jelly bean size of fresh feces of a cat, dog, sheep, cattle, hog, or horse
- Microscope slide and coverslip
- Small piece of newspaper
- Toothpick
- Tap water
- Eyedropper
- Gloves and eye protection

### IMPORTANCE TO INDUSTRY

The internal parasites with the most economic impact on livestock and dogs and cats include the roundworms (nematodes), flukes (trematodes), and tapeworms (cestodes). Heartworm (*Dirofilaria immitis*) is the leading cause of death in dogs.

### PROCEDURE

***Note:*** This exercise should be performed by the instructor as an observation lab.

#### For the Collection of the Fecal Sample

Fresh feces should be collected and used. At least several grams of fecal material should be gathered (about the size of a jelly bean).

1. Turn the plastic bag inside out.
2. Place your hand in the bag and pick up the fresh fecal material.
3. Pull the plastic bag over your hand so the fecal sample is now inside the bag and your hand is outside the bag.
4. Zip the bag closed.

#### Smear Technique

1. Place a clean microscope slide on a small piece of newspaper.
2. Put one drop of tap water on the center of the slide.
3. Using a toothpick, remove a small sample of the collected fecal material.
4. Mix the sample into the drop of water on the slide until the suspension is cloudy but the newsprint can be read.
5. Using the Fecal Sample Microscope Examination form, sketch and label one item from each of the fecal samples observed.

# FECAL SAMPLE MICROSCOPE EXAMINATION

**Name** ______________________ **Date** ______________________

| | | | |
|---|---|---|---|
| Label ________<br>Species ________ | Label ________<br>Species ________ | Label ________<br>Species ________ | Label ________<br>Species ________ |
| Label ________<br>Species ________ | Label ________<br>Species ________ | Label ________<br>Species ________ | Label ________<br>Species ________ |
| Label ________<br>Species ________ | Label ________<br>Species ________ | Label ________<br>Species ________ | Label ________<br>Species ________ |
| Label ________<br>Species ________ | Label ________<br>Species ________ | Label ________<br>Species ________ | Label ________<br>Species ________ |



Copyright © 2015 Cengage Learning®.

6. Remove any large bits of debris that may be present by scraping the toothpick horizontally over the cloudy solution (more drops of water may be needed).
7. Gently place the coverslip onto the slide.
8. Examine under the microscope. Begin with 10× power and then increase the power to see detail. When worm eggs and coccidial oocytes or other protozoa are present in the same sample, the smaller coccidian and giardia will float to the top toward the coverslip. When the power is 40× or higher, the eggs will be in focus, but the coccidian will not. The fine-adjustment knob will focus all three types of parasites.

## CONCLUSIONS

1. Which samples appeared to have evidence of parasites?

______________________________________________

______________________________________________

2. What treatment(s) would you provide these or this animal with?

______________________________________________

______________________________________________

## OPTIONAL INTERNET SUPPLEMENT

School of Veterinary Medicine- University of Wisconsin-Madison (http://www.vetmed.wisc.edu/pbs/vetpara/tutorial.html): This Web site provides additional information about internal parasites in livestock.

# EXERCISE 40

## It's All in the Genes—Understanding Genetics

### PURPOSE

This exercise provides students with an understanding of how different gene combinations result in different individuals.

### MATERIALS NEEDED

- Pencil or pen
- Calculator
- Text
- Internet access (optional)

### PROCEDURES

1. Study the following terms related to genetics:
   a. **heredity**—the passing on of traits or characteristics from parents to offspring
   b. **genes**—components of cells that determine the individual characteristics of living things
   c. **genetics**—the study of heredity
   d. **genotype**—what the genes look like
   e. **phenotype**—physical appearance of an individual
   f. **dominant**—gene that expresses itself to the exclusion of other genes
   g. **recessive**—gene that expresses itself only in the absence of a dominant gene
   h. **heterozygous**—pairs of genes that are different
   i. **homozygous**—pairs of genes that are alike
   j. **incomplete dominance**—neither gene expresses itself to the exclusion of the other
2. Complete the squares for the mating of the following two animals: Bb—Black bull and bb—Red cow Bb × bb (Observation 1)
3. Complete the squares for the mating of the following two animals: PP—Polled bull and pp—Horned cow PP × pp (Observation 2)
4. Complete the squares for the mating of the following two animals: BbPP—Black polled bull and bbpp—Red horned cow BbPP × bbpp (Observation 3)
5. Complete the squares for the mating of the following two animals: RR—Red Shorthorn bull and WW—White Shorthorn cow RR × WW (Observation 4)

## OBSERVATIONS

**1.** Complete the squares that follow:

COW

BULL

| | |
|---|---|
| | |
| | |



**2.** Complete the squares that follow:

COW

BULL

| | |
|---|---|
| | |
| | |



**3.** Complete the squares that follow:

COW

BULL

| | | | |
|---|---|---|---|
| | | | |
| | | | |
| | | | |
| | | | |



**4.** Complete the squares below:

COW

BULL

| | |
|---|---|
| | |
| | |

## CONCLUSIONS

**1.** Using the text or the Internet, define the following terms:

a. Dominant ______________________________

b. gene ______________________________

c. genetics ______________________________

d. genotype ______________________________

e. heredity ______________________________

f. heterozygous ______________________________

g. homozygous ______________________________

h. incomplete dominance ______________________________

i. phenotype ______________________________

j. recessive ______________________________

**2.** In Observation 1, identify the following:

a. The dominant characteristic ______________________________

b. The recessive characteristic ______________________________

c. The phenotypic ratio ______________________________

d. The genotypic ratio ______________________________

**3.** In Observation 2, identify the following:

a. The dominant characteristic ______________________________

b. The recessive characteristic ______________________________

c. The phenotypic ratio ______________________________

d. The genotypic ratio ______________________________

4. In Observation 3, identify the following:
   a. The dominant characteristic ____________________
   b. The recessive characteristic ____________________
   c. The phenotypic ratio ____________________
   d. The genotypic ratio ____________________
5. In Observation 4, identify the following:
   a. The dominant characteristic ____________________
   b. The recessive characteristic ____________________
   c. The phenotypic ratio ____________________
   d. The genotypic ratio ____________________

## OPTIONAL INTERNET SUPPLEMENT

Education Development Center, Enlivening Genetics Education, Web Labes (http://www2.edc.org/weblabs/). Web Lab on Punnett Squares.

This Web site is an interactive Punnett Square tutorial.

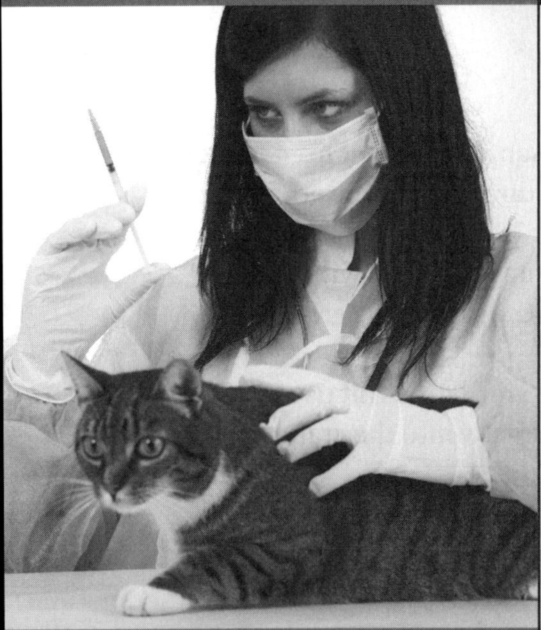
© photomak/Shutterstock.com

# EXERCISE 41

## This Will Hurt Only for a Minute—Injection Procedures

### PURPOSE

This lab enables students to practice procedures for properly vaccinating animals for disease prevention and control.

### MATERIALS NEEDED

- Two or three 10-mL syringes
- 1/2-inch needles
- 1-inch needles
- Three containers of water
- Food dye—red, green, blue
- Three or four grapefruit
- Ruler

### PROCEDURES

1. Study the following terms related to injection procedures.
   a. **injection**—the process of administering drugs by needle and syringe
   b. **antibiotic**—substance used to help prevent or control certain diseases of animals
   c. **intradermal**—between layer of skin
   d. **intramuscular**—in a muscle
   e. **intraperitoneal**—in the abdominal cavity
   f. **intrarumenal**—in the rumen
   g. **intravenous**—in a vein
   h. **subcutaneous**—under the skin
   i. **syringe**—an instrument used to give injections of medicine or to draw body fluids from animals
   j. **vaccination**—the injection of an agent into an animal to prevent disease
2. Add blue dye to one water container, red dye to a second water container, and green dye to the third. Add dye to the water until the water is a dark color.
3. Using the ½-inch needle, inject one grapefruit with 2 mL of green-colored dye directly under the skin of the grapefruit (subcutaneous).
4. Using a 1-inch needle, inject the same grapefruit used in Step 3 with 2 mL of red-colored dye directly into the grapefruit (intramuscular).
5. Using the ½-inch needle, inject one grapefruit with 5 mL of blue-colored dye directly under the skin of the grapefruit (subcutaneous).

6. Using a 1-inch needle, inject the same grapefruit used in Step 5 with 5 mL of green-colored dye directly into the grapefruit (intramuscular).
7. Using the ½-inch needle, inject one grapefruit with 10 mL of red-colored dye directly under the skin of the grapefruit (subcutaneous).
8. Using a 1-inch needle, inject the same grapefruit used in Step 7 with 10 mL of blue-colored dye directly into the grapefruit (intramuscular).
9. Wait 10 minutes then cut each grapefruit along a line where the intramuscular injection was given. Record your findings in Observation 1.
10. Peel the grapefruit around the area where the subcutaneous injection was given. Record your findings in Observation 2.

## OBSERVATIONS

1. Provide the information requested in the following for the intramuscular injection:

| Intramuscular Injection | Grapefruit 1 2 mL | Grapefruit 2 5 mL | Grapefruit 3 10 mL |
|---|---|---|---|
| Approximate Size of Area Colored with Dye | | | |



2. Provide the information requested in the following for the subcutaneous injection:

| Subcutaneous Injection | Grapefruit 1 2 mL | Grapefruit 2 5 mL | Grapefruit 3 10 mL |
|---|---|---|---|
| Approximate Size of Area Colored with Dye | | | |



## CONCLUSIONS

1. Which type of injection distributed the dye to the largest area?

_______________________________________________

_______________________________________________

**2.** Was the amount of area covered with the dye proportional from the smallest to the largest? Give reasons why it was or was not.

______________________________________________

______________________________________________

**3.** Which type of injection was easier for you to perform? Why?

______________________________________________

______________________________________________

**4.** Besides intramuscular and subcutaneous, list other types of injections that may be given.

______________________________________________

______________________________________________

## OPTIONAL INTERNET SUPPLEMENT

Dr. Raj Banga's Vet Video (http://www.vetvideos.com): Search-subcutaneous and intradermal injection techniques in cattle. USAID-Ethiopia Sheep and Goat Productivity Improvement Program (http://www.esgpip.org): Search—documents and publications/technical bulletins/giving injections to sheep and goats.

These Web sites are tutorials for the proper way to administer animal injections.

U.S. Fish and Wildlife Service

# EXERCISE 42

## Hot Wings Anyone?—Animal Anatomy and Physiology

### PURPOSE

This exercise enables students to practice manipulating dissecting equipment and to become familiar with organs and tissue found throughout the body of animals.

### MATERIALS NEEDED

- One raw chicken wing for each student (Chicken wings can be purchased at the local supermarket in the inexpensive family pack.)
- Scissors
- Paper plate
- Tweezers or forceps
- Protective gloves

### INTRODUCTION

A bird's wing is made up of groups of tissues and organs working together to perform a job. Before beginning the dissection, review the functions performed by the tissues and organs.

- Skeletal muscles are attached to bones.
- Skin is the membranous tissue that forms the outer covering of the body and provides a protective barrier from the outside environment.
- Muscle tissue is composed of bundles of skeletal muscle fibers. When these tissues expand and contract they produce motion in the wing.
- Fatty tissue, when stored on the underside of the skin, helps to keep the body warm, cushions and protects other body tissues, and stores vitamins A, D, E, and K.
- Blood vessels are the arteries, veins, and capillaries that transport blood throughout the body. Capillaries are too small to be seen without a microscope. Arteries have thicker walls than veins.
- Tendons are especially strong connective tissue that attaches skeletal muscle to bones.
- Cartilage helps prevent neighboring bones from grinding against each other.
- Bone provides structural support and manufactures red blood cells.
- Nerves are the bundles of fibers that transmit sensory stimuli and motion impulses.
- Ligaments are strong bands of connective tissue that connect two bones together.

### PROCEDURE

Read the description of each of the tissues or organs, and then begin the dissection for that particular tissue/organ. In the box provided, sketch an illustration of each tissue/organ listed in this lab.

**Skin**—The skin is the external covering of the entire wing. The skin will have a web like appearance between the bones. Look for evidence that the skin was covered with feathers. Cut a slit in the skin covering the largest bone and joint.

**Skin**

**Connective Tissue**—First lift a corner of the skin and, with tweezers or forceps, peel it back gently from the muscle. Notice the shiny, thin membrane that surrounds the muscle and attaches the muscle to the skin. This is connective tissue.

**Connective Tissue**

**Muscle Tissue**—Note the pink-orange bundles of fibers attached to the bone. This is muscle tissue.

**Muscle Tissue**

**Fatty Tissue**—Continue to peel back the skin slowly and gently until you locate a tissue between the skin and muscle that is yellow in color and greasy to the touch. If no fatty tissue can be found there, find a thick piece of skin and proceed to cut through it to see whether there is any fatty tissue. If the lab uses store-bought chicken wings, it is possible little or no fat can be observed.

**Fatty Tissue**

**Blood Vessels**—With the skin peeled back, scan the surface of the skeletal muscles for thin red tubes. These are blood vessels. Arteries and veins might also be located in bundles of connective tissue with nerves.

**Blood Vessels**

**Tendons**—Look at the top of the large bone. Notice the shiny white pieces of connective tissue that attach skeletal muscle to bone. These are tendons.

**Tendons**

**Cartilage**—Examine the top of the large bone. Locate a pearly white hard tissue found at the ends of the long bones. This is cartilage.

**Cartilage**

**Bone**—Cut through the muscle with your scissors to expose the hard white bone. Bones are hard (but not solid) tissue made up primarily of calcium and phosphorus. The scissors will not be able to cross-section the bone. Ask the instructor to cut the bone. Look into the cross section of the bone; the mass of red tissue is red marrow. It is the red marrow that manufactures red blood cells. Notice the bone surrounding the red marrow. The bone itself is not solid, but is an asymmetrical lattice work of calcium and phosphorus.

**Marrow**

**Cross Section of Bone**

**Nerves**—Nerves are usually found buried deep within an organism, often lying close to long bones. Sometimes an artery, vein, and nerve are held together in one bundle of connective tissue. Gently peel the muscle from the long bone and search carefully for a thin, white, threadlike tissue. This is a nerve. If this does not work, look for a bundle of connective tissue containing blood vessels. Gently pry it apart and search for the nerve.

**Nerves**

**Ligaments**—Cut through the skin and muscle down to the joint between the long bone and the center bone. Locate the strong white bands of connective tissue that connect the two bones together at the joint. This is a ligament.

**Ligaments**

## OPTIONAL INTERNET SUPPLEMENT

SchoolTube (http://www.schooltube.com): Search—chicken wing dissection.

This Web site walks students through a chicken wing dissection.

# SECTION 9

## Food Science and Technology

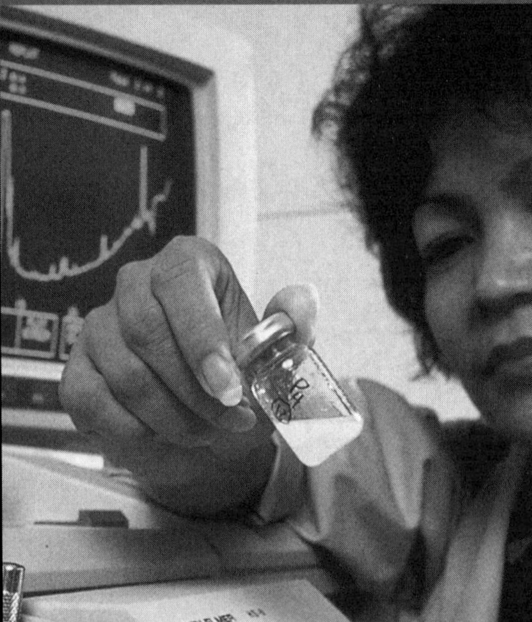
USDA/ARS #K-3512-3

# EXERCISE 43

## Break It on Down?—Food Nutrients

### PURPOSE

This exercise allows you to observe the different food nutrients found in the foods we eat.

### MATERIALS NEEDED

- Twenty test tubes
- Test tube rack
- Five eyedroppers
- Hot plate
- Sudan IV
- Biuret reagent
- Potassium iodide—iodine solution
- Benedict's solution
- Milk
- Cream or butter
- Syrup
- Canned whole corn
- Distilled water

### PROCEDURE

1. Group the test tubes in the test tube rack in groups of four. Label the five different groups as follows:
   a. Glucose—sugar
   b. Starch
   c. Fat
   d. Protein
   e. Water
2. Put 5 mL of water into each of the test tubes.
3. Add the following to the test tubes:
   a. Three to five drops of syrup to the four test tubes marked "Glucose."
   b. Three to five drops of juice from the canned corn to the four test tubes marked "Starch."
   c. Three to five drops of cream or melted butter to the four test tubes marked "Fat."
   d. Three to five drops of milk to the four test tubes marked "Protein."
   e. Three to five drops of distilled water to the four test tubes marked "Water."
4. Add 10 drops of Benedict's solution to the first tube of each group of four test tubes. Heat the four tubes gently in a hot water bath for 10 minutes. If sugar is present, the color will change to orange. Record your findings in the table in Observation 1.
5. Add one drop of potassium iodide—iodine solution— to the second tube in each group of four test tubes. Starch turns blue in the presence of iodine. Glycogen

(animal starch) turns mahogany red in the presence of iodine. Record your findings in the table in Observation 2.

6. Add 1 drop of Sudan IV to the third tube in each group of four test tubes. Sudan IV is an oil-soluble dye and will tag oil globules. Record your findings in the table in Observation 3.
7. Add 10 drops of Biuret reagent to the last test tube in each group of four test tubes. Biuret reagent changes from blue to violet pink if peptides (protein building blocks) are present. The color may fade. Allow 10 to 15 minutes for the reaction to take place. Record your findings in the table in Observation 4.

## OBSERVATIONS

1. Sugar test results

| Test Tube | Reaction (if any) |
|---|---|
| Syrup | |
| Corn juice | |
| Cream | |
| Milk | |
| Water | |



2. Starch test results

| Test Tube | Reaction (if any) |
|---|---|
| Syrup | |
| Corn juice | |
| Cream | |
| Milk | |
| Water | |

**3.** Fat test results

| Test Tube | Reaction (if any) |
|---|---|
| Syrup | |
| Corn juice | |
| Cream | |
| Milk | |
| Water | |



**4.** Protein test results

| Test Tube | Reaction (if any) |
|---|---|
| Syrup | |
| Corn juice | |
| Cream | |
| Milk | |
| Water | |



## CONCLUSIONS

**1.** Which foods reacted to the starch test?

______________________________

______________________________

**2.** Which foods showed the presence of sugar?

______________________________

______________________________

**3.** What were the protein foods?

______________________________

______________________________

**4.** Which foods reacted to the fat test?

**5.** What reactions took place in the distilled water?

**6.** List the foods (if any) that showed more than one of the food nutrients and which nutrients were present.

## OPTIONAL INTERNET SUPPLEMENT

Nutrition Analysis Tool (http://www.myfoodrecord.com/mainnat.html):

This web site allows students to input foods they have eaten and analyze them for nutrient content.

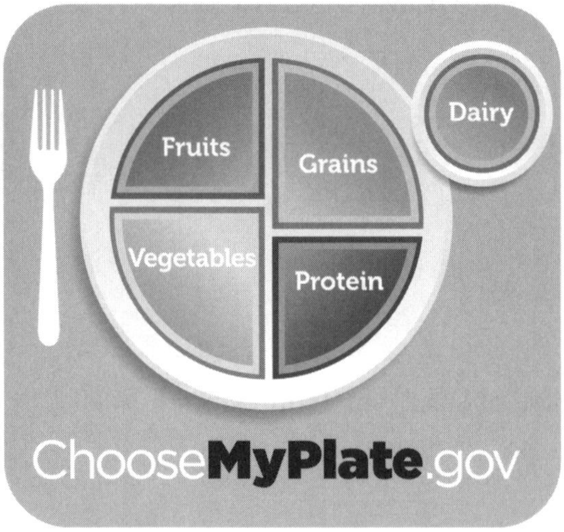


Courtesy of USDA. USDA does not endorse any products, services, or organizations.

# EXERCISE 44

## What's for Dinner?—Exploring Choose My Plate

### PURPOSE

Agriculture is the most important industry in the world. We could live without our laptops and ipods, but not without food. In this exercise, we will discover the kinds and amounts of food one person needs to live a healthy lifestyle.

### MATERIALS NEEDED

- Variety of foods from each food group
- Kitchen measuring tools
- Scale in ounces
- Internet access

### PROCEDURE

1. Log onto the Choose My Plate Web site (www.choosemyplate.gov).
2. To help you maintain a healthy body, it is important to know what foods and in what quantities you should eat. From the Choose My Plate home page, select the "Super Tracker" link. Create an account and fill out your personal information or that of a 16 year old, 160 lb. male that is 5'10", and is active 30–60 minutes a day.
3. Select "My Plan" and review the information provided. Fill out the chart in question number one of the conclusions sections.
4. For each food category, read the tips the web site provides to help individuals fill their daily nutritional need.
5. Using the foods provided by your instructor, measure out 10 ounces of food from the grains food group. Use at least three different kinds of grain-based foods to equal the total number of ounces your body needs. Keep in mind that half of the grains you consume throughout the day should come from whole grains.
6. Again, using the food provided by your instructor, measure the appropriate amount of food you should eat daily from each of the remaining food groups. Be sure to use at least two different food sources to represent each category.

## CONCLUSIONS

**1.** Fill out the chart below to reflect how much food from each category you should eat each day.

How many calories do you need for a healthy weight? ____________________

| Food Group | Amount |
|---|---|
| Grains | |
| Vegetables | |
| Fruit | |
| Milk | |
| Meat/Beans | |



**2.** For each food group, list two tips that will help you fill the daily recommendations.

______________________________________________________________________

______________________________________________________________________

**3.** What are whole grains? How can you tell if a food falls into this category?

______________________________________________________________________

______________________________________________________________________

**4.** What are the recommendations for oils and discretionary calories provided?

______________________________________________________________________

______________________________________________________________________

Courtesy of Price Chopper Supermarkets

# EXERCISE 45

## Is This Still Good?—Food Preservation

### PURPOSE

This exercise allows you to determine the presence of food additives in the foods we eat.

### MATERIALS NEEDED

- Bread
- Swiss cheese
- Hot plate
- Two 20-mL test tubes
- Two 250-mL beakers
- Distilled water
- Test tube holder
- Sulfuric acid solution
- Funnel
- Filter paper
- Spatula
- Ethyl alcohol
- Pencil
- Personal Safety Set

### PROCEDURE

1. Place 400 mL of distilled water into a 600 mL beaker and bring to a low boil on the hot plate.
2. Cut the Swiss cheese into small squares and place in a 250 mL beaker. Add 100 mL of distilled water.
3. Mash the cheese and stir with a spatula to mix the cheese and water as much as possible. Allow to sit while preparing the bread.
4. Place a slice of bread in a 250 mL beaker and add 100 mL of distilled water. Let the bread soak. Stir the mixture with a spatula to mix the bread and water as much as possible.
5. Using the funnel with filter paper, filter approximately 10 mL of the cheese–water mixture in one test tube.
6. Follow the same procedure for the bread mixture.
7. Add 5 mL of sulfuric acid solution to both the Swiss cheese and bread solutions. Record the reaction in Observation 1.
8. Add 3 mL of ethyl alcohol to each test tube and place in water bath on hot plate.
9. When the solutions begin to boil, carefully observe the odor coming from the vapors. In Observation 2, record the odor you detect.

***Note:*** An odor similar to pineapple indicates ethyl propionates, that are mold inhibitors.

## OBSERVATIONS

1. What happened to the cheese solution when the sulfuric acid was added?

2. What odor did you detect when you heated the cheese solution? What odor did you smell when you heated the bread solution?

## CONCLUSIONS

1. Why are preservatives necessary in foods such as cheese and bread?

2. What are some natural preservatives found in food?

3. If food preservatives were not used, how do you think it would affect food supply and prices?

## OPTIONAL INTERNET SUPPLEMENT

Food Additives and Ingredients Association, Web page on Understanding Food Additives (http://www.understandingfoodadditives.org/):

This web site explains food preservation techniques and food additives.

USDA

# EXERCISE 46

## How Cheesy—Dairy Product Production

### PURPOSE

The purpose of this example is to demonstrate the process involved in producing cheese.

### MATERIALS NEEDED

- 12-inch piece of cheesecloth
- Hot plate
- Plastic cups—16 ounces
- each
- Cooking thermometer
- Milk
- Salt
- Rennin or rennilase

### PROCEDURE

1. Heat the milk to 88° to 90° F on the hot plate. Use the thermometer to monitor the temperature.
2. Fill the plastic cups approximately two-thirds to three-fourths full.
3. Immediately add 12 drops of the rennin solution to the milk. Stir continuously until the rennin enzyme is completely mixed in.
4. Let the milk set undisturbed until the curd (solid portion) has separated from the whey (liquid portion).
5. Filter the liquid from the solid by folding the cheesecloth in half and placing over the top of the cup to act as a strainer.
6. Pour the liquid into another plastic cup.
7. Salt the cheese (curds) and allow it to further harden.
8. Cut and eat.

## OBSERVATIONS

1. Describe the color of the curds. Describe the color of the whey.

2. Describe the taste of the finished cheese.

## CONCLUSIONS

1. What function does the enzyme rennin serve in this experiment?

2. How do you explain the different tastes of different cheeses?

## OPTIONAL INTERNET SUPPLEMENT

University of Cincinnati-Clermont College Biology Department (http://biology.clc.uc.edu/fankhauser/): Search—cheese page/ cheese, basic hard, 5 gallon.

This web site walks you through the cheese making process.

© iStockphoto/Jason Lugo

# EXERCISE 47

## Does This Milk Make Me Look Fat?—Homogenization

### PURPOSE

Before milk hits the shelves of your local grocery store, it is homogenized. In this process, the fat in the milk is broken up to prevent it from separating. In this exercise, you will be able to see the difference between homogenized milk and raw milk.

### MATERIALS NEEDED

- Compound microscope with an oil immersion lens
- Microscope slide and coverslip
- Raw milk from a dairy or a health food store that carries raw milk
- Homogenized milk
- Internet access
- Methylene blue

### PROCEDURE

1. Put a drop of raw milk on one slide and a drop of homogenized milk on another slide. Add a drop of methylene blue to each drop of milk to make viewing under a microscope easier.
2. Place a coverslip over each sample.
3. With the oil immersion lens, look at the differences in the size of the fat cells.

### OBSERVATIONS

Sketch the fat globules of the homogenized milk sample and the fat globules from the raw milk sample.

| **Homogenized fat globules** | **Raw fat globules** |
| --- | --- |

## CONCLUSIONS

1. Describe the differences between the two milk samples.

_______________________________________________

_______________________________________________

2. Using the text, define homogenization.

_______________________________________________

_______________________________________________

3. Researching the process involved in homogenization, give a brief description of your findings.

_______________________________________________

_______________________________________________

## OPTIONAL INTERNET SUPPLEMENT

Ohio State University- Department of Food Science and Technology (http://www.osu.edu/): Search—homogenization of milk fat globules.

This web site provides information about milk homogenization.

USDA/ARS #K-48191

# EXERCISE 48

## Do You Have a Sweet Tooth?—Fruit Sugars

### PURPOSE

This exercise outlines how to measure sugar levels in fruits.

### MATERIALS NEEDED

- Refractometer
- A variety of fruits
- Kitchen knife
- Blender (optional)

### PROCEDURE

1. Obtain a drop of liquid (juice) from a piece of fruit. This can be done by squeezing or mixing it up in a blender. Open the plastic top of the refractometer and drop a bead of the juice on the glass window.
2. Close the plastic top.
3. Hold the refractometer so the light source (outside sun is best) is directed toward the plastic top, not the end of the refractometer.
4. Read the degree brix at the point where the white and blue colors meet.
5. Record the degree brix in the chart on the following page.
6. Repeat the steps above for all of the fruit samples.

### OBSERVATIONS

Before measuring the actual brix, list the fruits used in this lab in the left-column. Then list the fruits in the order in which you *think* they rank in sweetness from highest to lowest.

| Fruits | Ranking from Highest to Lowest in Sweetness |
|---|---|
| ____________________ | ____________________ |
| ____________________ | ____________________ |
| ____________________ | ____________________ |

**Percent Sugar or Degree Brix**

| Fruit | Brix | Comments |
|---|---|---|
| | | |
| | | |
| | | |
| | | |



## CONCLUSIONS

1. What does a refractometer actually measure?

2. What is the name of the sugar found in fruits?

3. Of your favorite fruits, which do you think is the sweetest?

## OPTIONAL INTERNET SUPPLEMENT

USDA, ARS, Products and Services, Nutrient Data, SR25- Reports by single nutrients (http://www.ars.usda.gov/Services/docs.htm?docid=22769): Search—total sugars.

This web site is a data base of foods listed by sugar content. Select a few items and compare sugar content.

# SECTION 10

# Communications and Management in Agriscience

# EXERCISE 49

## Where Did the Money Go?—Understanding Diminishing Returns

### PURPOSE

This exercise will provide an understanding of the law of diminishing returns.

### MATERIALS NEEDED

- Pencil
- Calculator

### PROCEDURE

**1.** Study the terms associated with diminishing returns:

a. **marginal**—refers to incremental changes, increases or decreases, that occur at the edge or margin (additional)

b. Δ—shorthand for "change in"

c. **total physical product (TPP)**—the total output or yield

d. **average physical product**—average amount of output produced by each unit of input at each input level:

$$\text{APP} = \frac{\text{Total physical product (TPP)}}{\text{Input level}}$$

e. **marginal physical product (MPP)**—the additional or extra total physical product produced by using an extra unit of input:

$$\text{MPP} = \frac{\Delta\ \text{total physical product}}{\Delta\ \text{input level}}$$

f. **marginal value of product (MVP)**—the additional or marginal income received from using an additional unit of input:

$$\text{MVP} = \frac{\Delta\ \text{total value of product (TVP)}}{\Delta\ \text{input level}}$$

g. **total value of product (TVP)**—the product of multiplying quantity of output (TPP) by selling price.

h. **marginal input cost (MIC)**—the change in total input cost or the addition to total input cost caused by using an additional unit of input

$$\text{MIC} = \frac{\Delta \text{ total input cost}}{\Delta \text{ input level}}$$

i. The **profit maximizing point** is the input level where MVP = MIC.

**2.** Study the following examples:

Input price = $10.00 per unit
Output price = $2.50 per unit

| Input Level | TPP | APP | MPP | TVP ($) | MVP ($) | MIC ($) |
|---|---|---|---|---|---|---|
| 0 | 0 | – | – | 0 | – | – |
| 1 | 10 | 10 | 10 | 25 | 25 | 10 |
| 2 | 28 | 14 | 18 | 70 | 45 | 10 |
| 3 | 42 | 14 | 14 | 105 | 35 | 10 |
| 4 | 52 | 13 | 10 | 130 | 30 | 10 |
| 5 | 60 | 12 | 8 | 150 | 20 | 10 |
| 6 | 66 | 11 | 6 | 165 | 15 | 10 |
| 7 | 70 | 10 | 4 | 175 | 10 | 10 |
| 8 | 72 | 9 | 2 | 180 | 5 | 10 |
| 9 | 70 | 7.78 | −2 | 175 | −5 | 10 |
| 10 | 68 | 6.80 | −4 | 170 | −10 | 10 |



a. Input level and total physical product (TPP) are given. Input price and output price are given.

b. Study the **a**verage **p**hysical **p**roduct (APP) column. Examples:

$$\text{APP} = \frac{\text{TPP}}{\text{input}} = \frac{10}{1} = 10 \text{ for level 1}$$

$$\text{APP} = \frac{\text{TPP}}{\text{input}} = \frac{28}{2} = 14 \text{ for level 2}$$

c. Study the **m**arginal **p**hysical **p**roduct (MPP) column.
Examples:

$$MPP = \frac{TPP}{\Delta\ \text{input}} = \frac{(10 - 0)}{(1 - 0)} = 10 \text{ for level 1}$$

$$MPP = \frac{TPP}{\Delta\ \text{input}} = \frac{(28 - 10)}{(2 - 1)} = 18 \text{ for level 2}$$

d. Study the **t**otal **v**alue of **p**roduct (TVP) column.
Examples:

TVP = TPP × selling price = 10 × \$2.50
= \$25.00 for level 1

TVP = TPP × selling price = 28 × \$2.50
= \$70.00 for level 2

e. Study the **m**arginal **v**alue of **p**roduct (MVP) column.
Examples:

$$MVP = \frac{TVP}{\Delta\ \text{input}} = \frac{(25 - 0)}{(1 - 0)} = \$25.00 \text{ for level 1}$$

$$MVP = \frac{TVP}{\Delta\ \text{input}} = \frac{(70 - 25)}{(2 - 1)} = \$45.00 \text{ for level 2}$$

f. Study the **m**arginal **i**nput **c**ost (MIC) column.
Examples:

$$MIC = \frac{\text{total input cost}}{\Delta\ \text{input}} = \frac{(1 - 0)(\$10.00)}{(1 - 0)} = \frac{10}{1} = 10$$

$$MIC = \frac{\text{total input cost}}{\Delta\ \text{input}} = \frac{(2 - 1)(\$10.00)}{(2 - 1)} = \frac{10}{1} = 10$$

g. The profit maximizing point is where MVP = MIC. If you look at input level 7, you find MVP = 10 and MIC = 10.

h. The law of diminishing returns states as additional units of a variable input are used in combination with one or more fixed inputs, the marginal physical product will eventually begin to decline. Study the MVP column on the chart.

## OBSERVATIONS

Complete the following chart.
Fertilizer costs $15.00 per unit.
Corn sells for $3.00 per bushel.

| Units of Fertilizer | TPP (Corn) | APP | MPP | TVP ($) | MVP ($) | MIC ($) |
|---|---|---|---|---|---|---|
| 0 | 70 | – | – | 0 | – | – |
| 1 | 96 | | | | | |
| 2 | 118 | | | | | |
| 3 | 129 | | | | | |
| 4 | 136 | | | | | |
| 5 | 140 | | | | | |
| 6 | 150 | | | | | |
| 7 | 140 | | | | | |
| 8 | 136 | | | | | |



1. Complete the average physical product (APP) column.
2. Complete the marginal physical product (MPP) column.
3. Complete the total value of product (TVP) column.
4. Complete the marginal value of product (MVP) column.
5. Complete the marginal input cost (MIC) column.

## CONCLUSIONS

1. What is the profit-maximizing amount of fertilizer to use on corn?

2. If the TVP column is higher on level 6, why is it not the profit-maximizing level?

**3.** If fertilizer cost only $12.00 per unit, what would be the profit-maximizing level of fertilizer to use? Why?

## OPTIONAL INTERNET SUPPLEMENT

Agri Info (http://www.agriinfo.in/): Search—The Law of Diminishing marginal returns.

This web site describes diminishing returns in agriculture.

# EXERCISE 50

## I Owe You—Interest and Credit

### PURPOSE

Farm equipment is expensive. Loans are often needed to purchase it. This exercise demonstrates the different methods of charging interest when borrowing money.

### MATERIALS NEEDED

- Pencil
- Calculator

### PROCEDURE

**1.** Study the following terms:

a. **principal**—the amount of money received from a loan transaction

b. **interest**—the charge paid for use of money borrowed from an institution or person

c. **interest rate**—the charge paid for money borrowed, expressed as a percentage of the principal

d. **time**—the length of payback for money borrowed, usually expressed in terms of years

e. **true** or **actual annual rate**—a standard or basic interest rate for comparison of different methods of charging interest

f. **simple interest**—the interest rate on a loan with a single (one) payment

g. **interest on the unpaid balance**—a method of charging interest causes the interest payments to decline as the principal loan balance declines

h. **discounted interest**—a type of loan that requires the interest to be paid in advance or at the time the loan is received. The interest is deducted from the original amount of the loan, leaving the borrower with less than the amount actually borrowed

i. **add-on-interest**—usually charged on installment loans with interest charged on the full amount of the loan for the entire time of the loan

**2.** Study the following examples:

### Simple Interest

a. \$12,000 is borrowed for one year at 10 percent interest.

**Formula:** $I = PRT$

$I = \$12{,}000 \times 10\% \times 1\text{ (year)} = \$1{,}200$

The simple interest rate is also the true annual rate: 10 percent.

### Interest on Unpaid Balance

b. \$12,000 is borrowed at 10 percent interest, \$6,000 will be repaid at the end of 6 months, and the remaining \$6,000 will be paid at the end of the year.

**Formula:** $I = PRT$

***Note:*** The equation above will be applied twice. First for the total principal and then again for the unpaid principal amount after a large payment has been made. The two interest costs are then added together as shown here.

$$I = \$12{,}000 \times 10\% \times \frac{1}{2}\text{ (year)} = \$600$$

$$I = \$6{,}000 \times 10\% \times \frac{1}{2}\text{ (year)} = \frac{\$300}{\$900}$$

When interest is paid only on the unpaid balance, the stated interest rate is the same as the true annual rate: 10 percent.

### Discounted Interest

c. \$12,000 is borrowed at 10 percent but is discounted so \$1,200 interest is deducted.

**Formula:**

$$\text{rate} = \frac{\text{Interest}}{\text{Adjusted principal} \times \text{time}} \times 100$$

$$= \frac{\$1{,}200}{(\$12{,}000 - 1{,}200) \times 1}$$

$$= \frac{\$1{,}200}{\$10{,}800 \times 1}$$

$$= 11.11\%$$

### Add-on Interest

d. \$12,000 is borrowed at 10 percent for 3 years with 36 equal payments.

$$I = PRT = \$12{,}000 \times 10\% \times 3\text{ (years)} = \$3{,}600$$

**Formula:** $$\text{true rate} = \frac{2 \times \text{no. of payment} \times \text{interest charged}}{\text{Principal} \times \text{time} \times (\text{no. of payments} + 1)}$$

$$= \frac{2 \times 36 \times 3{,}600}{12{,}000 \times 3 \times 37}$$

$$= \frac{259{,}200}{1{,}332{,}000}$$

$$= 19.46\%$$

The following chart shows the difference in interest paid and true annual interest rate for different types of interest charged on the following.
Amount borrowed = $12,000
Interest rate = 10%

| Kind of Interest | Rate | Time (Years) | Interest Paid | True Annual Interest Rate |
|---|---|---|---|---|
| Simple Interest | 10% | 1 | $1,200 | 10% |
| Interest on Unpaid Balance | 10% | 1 | $900 | 10% |
| Discounted Interest | 10% | 1 | $1,200 | 11.11% |
| Add-on Interest | 10% | 3 | $3,600 | 19.46% |



## OBSERVATIONS

**1.** You are planning to borrow $15,000. You have been offered a 12 percent rate with the following options:

a. Simple interest for one year.
b. Interest on unpaid balance for one year with half due in 6 months and half due at the end of the year.
c. A discounted loan for one year.
d. Add-on-interest for three years with 36 payments.
Calculate the amount of interest and true APR for each of the preceding options and record them in the following chart.

| Kind of Interest | Rate | Time (Years) | Interest Paid | True Annual Interest Rate |
|---|---|---|---|---|
| Simple Interest | 12% | 1 | | |
| Interest on Unpaid Balance | 12% | 1 | | |
| Discounted Interest | 12% | 1 | | |
| Add-on Interest | 12% | 3 | | |



## CONCLUSIONS

1. What is the lowest interest rate? What is the highest?

2. If you were to make payments each month or put in the bank an amount equal to a payment (for those due in one lump sum), which would cost the most per month? Which would cost the least?

3. Based on the answer to conclusion 2, why do you think some people would choose an add-on-interest loan?

## OPTIONAL INTERNET SUPPLEMENT

United States Department of Agriculture (http://www.usda.gov): Search—agriculture grants and loans.

This web site provides information for government grants and loans.

© NAN728/Shutterstock.com

# EXERCISE 51

## In the Money—Writing a Business Plan

### PURPOSE

Starting a new business is a complicated process and must begin with a plan. In this exercise, you will practice writing a simple business plan.

### MATERIALS NEEDED

- Computer
- Internet access
- Calculator

### PROCEDURE

In this exercise, you will be creating a business plan for a business of your choice. If you currently are involved in an agribusiness, use that information for your business plan. If not, use your imagination to create one. As you answer the questions below, estimate any dollar amounts you need. The purpose of this exercise is to understand the process involved in the planning, not focusing on actual dollar amounts.

1. Determine what product or service your business will provide.
2. Who is your target customer?
3. What makes your product special and you committed to the business?
4. Estimate what kinds of permits, insurance, location costs, loan payments and other expenses you may have.

## OBSERVATIONS

1. Fill out the estimated expense table below.

| Expense | Cost/Month |
|---|---|
| | |
| | |
| | |
| | |
| | |
| | |
| | |



2. Add your costs. ______________________________

3. Taking costs into consideration, how much will you charge for the product or service?

______________________________

## CONCLUSIONS

Write a one page paper describing your plan.

1. Include an introduction that describes your business.
2. In the body of the paper, describe your target customer and why they would want or need your service or product.
3. Describe what makes your product unique and why you are personally committed to your business.
4. Briefly explain the following; your estimated costs, what you will need to charge for your product or service, and how much you will have to sell to make a profit.
5. Write a summary of the information you provided in the context of the paper. Be sure to include the information that shows your business can be profitable.

## OPTIONAL INTERNET SUPPLEMENT

AgPlan (https://www.agplan.umn.edu/):

This web site is a free interactive tool for writing an agricultural business plan.

Courtesy of FFA

# EXERCISE 52

## It's a Balancing Act—Balance Sheets

### PURPOSE

This exercise demonstrates the use of balance sheets in agribusiness.

### MATERIALS NEEDED

- Pencil
- Calculator

### PROCEDURE

1. Study the following terms:
   a. **balance sheet**—a form showing the financial condition of a business at a definite point in time. It lists all assets, value of assets, and liabilities of the business. It is also known as a net worth statement, financial statement, or statement of financial condition.
   b. **assets**—the property or resources owned and controlled by a business.
   c. **current assets**—cash or other assets that can be converted to cash within 12 months. Examples: checking and/or savings accounts; accounts receivable; inventory held for sale (market animals and stored crops); other near-cash items such as securities, stocks and bonds; and cash value of life insurance.
   d. **intermediate assets**—resources or production items with useful lifetimes of 1 to 10 years. Examples: equipment, machinery, and breeding stock. These assets are generally depreciable.
   e. **fixed assets**—also called long-term assets; include permanent items such as real estate and improvements on buildings. The useful lifetime is generally over 10 years.
   f. **liabilities**—all the debt obligations of the business
   g. **current liabilities**—debts due within the operating year, that is, a 12-month period. Examples: notes and accounts payable, rents, taxes, interest, and principal payments due on intermediate or long-term debt within the next 12 months.

h. **intermediate liabilities**—nonreal-estate debt that corresponds to intermediate assets. Loan terms are normally for a period of 1 to 10 years. Examples: improvements to real estate, equipment purchases, breeding livestock and dairy stock, and capital requirements for major adjustments in farm operation.
i. **long-term liabilities**—mortgages and land contracts on real estate minus principal due within 12 months
j. **net worth**—total assets minus total liabilities; also called owner's equity
k. **liquidity**—the ability of a business to make enough cash to pay bills without disrupting business. One of the best measures of liquidity is current ratio.
l. **current ratio** $= \dfrac{\text{Current assets}}{\text{Current liabilities}}$
m. **intermediate ratio**—the intermediate liquidity of the business can be determined by the intermediate ratio.

$$\text{intermediate ratio} = \frac{\text{Currewnt + intermediate assets}}{\text{Current + intermediate liabilities}}$$

n. **solvency**—the ability to pay debts; shows the amount that would be left after all assets are converted to cash and all debts paid. Determined by:

$$\text{net capital ratio} = \frac{\text{Total asset}}{\text{Total liabilities}}$$

o. **equity**—the owner's share of the business. Shown by debt-equity ratio; also known as the debt-to-net worth ratio.

$$\text{(debt-to-net worth ratio)} = \frac{\text{Total liabilities}}{\text{Net worth (owner's equity)}}$$

Lenders are interested in this ratio because they usually prefer to provide loans that are equal to less than 50 percent of assets. Lenders prefer a debt-equity ratio of less than 1.

**2.** Using the following information,

| | |
|---|---|
| Current assets =$25,000 | Current liabilities =$15,000 |
| Intermediate assets =$75,000 | Intermediate liabilities =$50,000 |
| Fixed assets =$200,000 | Long-term liabilities =$150,000 |

a. Figure the current ratio.
b. Determine the intermediate ratio.
c. Use the net capital formula to determine the net capital ratio.
d. Determine the net worth.
e. Determine the debt-equity ratio.

# OBSERVATIONS

**1.** Use the following information to complete the accompanying balance sheet.

| | |
|---|---|
| Value of machinery | $75,000.00 |
| Loan on feeder steers | $18,000.00 |
| Value of cow herd | $21,000.00 |
| Mortgage on land | $125,000.00 |
| Checking account balance | $2,700.00 |
| Value of land | $337,000.00 |
| Bank note due next month | $16,500.00 |
| Balance on machinery loan | $34,000.00 |
| Value of feeder steers | $29,000.00 |
| Value of stored grain | $23,00.00 |

a. Fill in the current assets.
b. Fill in the current liabilities.
c. Complete the intermediate assets.
d. Complete the intermediate liabilities.
e. What are the fixed assets? List them on the balance sheet.
f. Add the long-term liabilities to the sheet.
g. Determine the net worth.

**BALANCE SHEET**

| | |
|---|---|
| **Current Assets** | **Current Liabilities** |
| ______________________ | ______________________ |
| ______________________ | ______________________ |
| **Intermediate Assets** | **Intermediate Liabilities** |
| ______________________ | ______________________ |
| ______________________ | ______________________ |
| **Fixed Assets** | **Long-Term Liabilities** |
| ______________________ | ______________________ |
| **Total Assets** ______________ | **Total Liabilities** ______________ |
| **Net Worth** ______________ | |

## CONCLUSIONS

1. What is the current ratio?

2. What is the intermediate ratio?

3. What is the net capital ratio?

4. What is the debt-equity ratio?

5. Is the business solvent?

## OPTIONAL INTERNET SUPPLEMENT

Penn State Department of Agriculture Science (http://agsci.psu.edu/): Search—financial guidelines for agriculture producers.

This web site is a financial guide for agricultural businesses.

© michaeljung/Shutterstock.com

# EXERCISE 53

## Rich Man, Poor Man—Profit–Loss Statements

### PURPOSE

The purpose of this exercise is to demonstrate the use of profit–loss statements in agribusiness.

### MATERIALS NEEDED

- Pencil
- Paper
- Calculator

## PROCEDURE

**1.** Study the following terms:

a. **profit and loss (income) statement**—the financial record that reflects the profitability of the business over a specified period of time

b. **income**—cash income includes money received from the sale of crops, livestock, and livestock products during the year. It also includes custom work and government payments. Gain from the sale of capital assets is also entered as cash income; the entry is the difference between the selling price and the book value. Non-cash income includes products used in the home or given to a farm for custom work. Positive changes in inventory are also considered non-cash income.

c. **expenses**—cash expenses are monies paid to operate the business, such as those for seed, feed, fertilizer, market livestock, and fuel. Fixed-cash expenses include taxes; interest; insurance; and, in some cases, repairs. Non-cash expenses include depreciation on machinery, equipment, buildings, and purchased breeding stock. Negative changes in inventory are recorded as non-cash expenses. Not recorded as expenses are costs of any new capital assets and principal payments on loans.

d. **cash accounting method**—income and expenses are recorded only when actual cash transactions occur. Most non-cash income and expenses are not recorded under the cash method. Depreciation is an exception, as is the value of farm products used in the home. Both of these are included under the cash method.

e. **accrual method**—records income when it is earned and expenses when they are incurred. Inventory changes are included in the accrual method. Because of this, it is actually a more accurate method of accounting

f. **net cash income**—the difference between total cash income and total cash expenses

g. **net farm income**—net cash income plus non-cash adjustments

h. **return to capital** or **return on investment**—a measure of profitability based on a ratio obtained by dividing the return to total capital by total farm assets

i. **return to capital**—adjusted net farm income minus opportunity cost of labor minus opportunity cost of management

j. **ratio of return to capital (%)** $= \frac{\text{Return to total capital}}{\text{Total farm assets}} \times 100$

k. **adjusted net farm income**—net farm income plus interest paid minus value of unpaid family labor

l. **return to labor and management**—that portion of net farm income that remains to pay the owner for personal labor and management after capital is paid a return equal to its opportunity cost. Return to labor and management equals adjusted net farm income minus opportunity cost on total capital.

m. **return to labor**—return to labor and management minus opportunity cost of management

n. **return to management**—return to labor and management minus opportunity cost of labor

o. **return to equity**—net farm income minus opportunity cost of own labor minus opportunity cost of management minus value of unpaid family labor

p. **rate of return to equity (%)** $= \frac{\text{Return to equity}}{\text{Net worth}} \times 100$

**2.** Working with formulas:

| | |
|---|---|
| Net farm income | $80,000.00 |
| Value of operator's management | $5,000.00 |
| Owner's equity | $500,000.00 |
| Interest paid | $25,000.00 |
| Value of total assets | $750,000.00 |
| Value of operator's labor | $15,000.00 |
| Opportunity cost of capital | 10% |

a. What is the return to capital in percent?____________
b. What is the return to management?____________
c. What is the return to equity in percent?____________

## OBSERVATIONS

**1.** Use the following information to complete the profit and loss statement:

| | |
|---|---|
| Grain sorghum sales | $30,000.00 |
| Alfalfa hay sales | $9,500.00 |
| Building depreciation | $2,800.00 |
| Increase in inventory of livestock | $15,000.00 |
| Feeder calf sales | $41,000.00 |
| Total farm cash operating expenses | $57,000.00 |
| Machinery depreciation | $12,000.00 |
| Decrease in crop inventory | $4,000.00 |

## CONCLUSIONS

**1.** What is the total revenue?

____________

____________

____________

**2.** What are the total expenses?

____________

____________

____________

**3.** What is the net farm income?

____________

____________

____________

**4.** What efficiency factors can be determined using the information from a profit–loss statement?

______________________________________________________________

______________________________________________________________

______________________________________________________________

## OPTIONAL INTERNET SUPPLEMENT

Iowa State University Extension (http://www.extension.iastate.edu): Search—Farm Financial Statements + C3-56.

This web site offers free financial work sheets.

# EXERCISE 54

## Money, Money, Money—Interest

### PURPOSE

Interest is calculated in different ways in the business world. The purpose of this lab is to learn how different types of interest are calculated.

### MATERIALS NEEDED

- Pencil
- Paper
- Calculator

### PROCEDURE

**1.** The formula for calculating basic simple interest is as follows;

$I$ = PRT
$I$ = Total interest
$P$ = Principal (loan amount)
$R$ = Interest rate as a decimal
$T$ = time

Example:
A loan for $1,000 is taken at 10 percent interest to be paid at the end of 5 years. What is the total amount of interest to be paid?

$I$ = (1,000)(.10)(5)
$I$ = $500

**2.** The formula to calculate the future value of money with simple compound interest is as follows:

$\text{FV} = \text{PV} \times (1 + r)^n$
FV = Future value of money
PV = Present Value (at start)
$r$ = interest rate as a decimal
$n$ = number of periods (usually years)

Example:
You invest $500 in a money market at a 6 percent interest rate. What will the future value be in 5 years?

$\text{FV} = 500 \times (1 + .06)^5$
FV = $669.11

3. The formula to calculate compound interest is as follows:

$$A = P\left(1 + \frac{r}{n}\right)^{nt}$$

$P$ = principal amount (initial investment)
$r$ = annual interest rate (as a decimal)
$n$ = number of times the interest is compounded per year
$t$ = number of years
$A$ = amount after time (t)

Example:
A loan is taken out for 10,000 with an interest rate of 3 percent that is compounded quarterly ( four times a year). How much interest will be paid at the end of the loan in 5 years?

$A = 10{,}000(1 + .03/4)^{4 \times 5}$
$A = 10{,}000(1.0075)^{20}$
$A = 10{,}000(1.161184)$
$A = 11{,}611.84$
$10{,}000 - 11{,}611.84 = \$1{,}611.84$ total interest paid

## OBSERVATIONS

1. Jake took out a basic simple interest loan for $1,000.00 at a 6 percent. Using the formula given in Procedure #1, figure how much interest he will pay by the time the loan is paid off in five years.

_______________

_______________

2. Using the simple compound interest formula from Procedure #2, calculate the future value (total amount he will need to repay) on Jake's loan if he was given a simple compound loan instead of a basic simple interest loan.

_______________

_______________

3. Using the compound interest formula given in Procedure #3, figure the amount of interest Jake would have to pay if he took out a compound loan.

_______________

_______________

## CONCLUSIONS

**1.** How much more interest would Jake have paid if he were to have had a compound interest loan instead of the basic interest loan?

______________________________________________

______________________________________________

**2.** When considering barrowing money, why is it important to understand what type of interest will be applied to a loan?

______________________________________________

______________________________________________

## OPTIONAL INTERNET SUPPLEMENT

The Web sites below are online interest calculators for simple, simple compound and compound interest. These sites can be used to check your work. Discovery Education's Web Math Web site (http://www.webmath.com/): Use the quick jump tool to navigate to "Interest, Simple" or "Interest, Compound." Time Value Software (http://www.timevalue.com/): Search—products/ Tcalc/ overview / investments/ what will my investment be worth in the future?

© Yeko Photo Studio/Shutterstock.com

# EXERCISE 55

## Can't We All Just Get Along—Cooperative Math Simulations

### PURPOSE

In this exercise, you will use basic math skills employed in cooperative businesses.

### MATERIALS NEEDED

- Pencil
- Paper
- Calculator
- Textbook

### PROCEDURE

**1.** In the United States, 2,000 marketing cooperatives serve 24,000,000 farmers for a total volume of $10 million. On the average, what dollar amount did each cooperative contribute to the total volume? __________________

__________________________________________

**2.** Paul Stark did $2,000 worth of business with his supply cooperative last year. At the end of the fiscal year, Paul received a $70 patronage refund. What percent of his purchases did the refund represent? __________________

__________________________________________

**3.** You are a member of a marketing cooperative with a membership of 920. The cooperative by-laws provide one vote per membership plus an additional vote for each 10 tons of fruit marketed. Assuming all members voted, 2,500 votes were tallied, and you marketed 5 percent of the total tonnage marketed, how many votes were you allowed to cast? __________________

__________________________________________

**4.** An egg marketing cooperative withholds $.02 per dozen for its revolving fund. A member is credited with $2,345 in the fund during 2002. If his production average was 175 eggs per hen, how many laying hens did he have in his flock? __________________

__________________________________________

5. An egg marketing co-op sells its eggs in cases of 30 dozen. If the co-op withheld $.01 cent per dozen for the revolving fund and it credited a member with $2,200, how many 30-dozen cases did the member market in that year? ______________

______________________________

6. James Dillon did $7,500 in business with his supply co-op last year. At the end of the year, he received a patronage refund in the amount of $316. What percent of his year's purchases did this cash refund represent? ______________

______________________________

7. A member of a fruit-marketing co-op delivered 900 tons of fruit to his co-op. Sixty-five percent was sold as fresh fruit, netting $150 per ton; 30 percent was sold for canning, netting $125 per ton; and 5 percent was sold for drying, netting $85 per ton. If the co-op retains 4 percent of all sales for its revolving fund, what did the member receive in cash for the season? ______________

______________________________

## CONCLUSIONS

1. What are the benefits of a cooperative business? ______________

______________________________

______________________________

2. If you were interested in starting a co-op business, how would you go about recruiting others to join? ______________

______________________________

______________________________

## OPTIONAL INTERNET SUPPLEMENT

United States Department of Agriculture (http://www.usda.gov): Search—Cooperative Benefits and Limitations.

This web page discusses the advantages and disadvantages of co-ops.